CW00585650

Green Gold

The Story of Lebanese Olive Oil

Turning Point

Green Gold

The Story of Lebanese Olive Oil

Written by
Sabina Mahfoud
Photography by
Roger Moukarzel
Recipes by
Mary Elizabeth Sabieh
Food Photography by
Kamal Mouzawak

Turning Point

Editors: Warren Singh Bartlett and Faerlie Wilson
Design: Maya Tawil
Printer: Dots (Dar el-Kotob)

ISBN: 9953-0-0026-3

This edition first published 2007

Published by: Turning Point
15th Floor
Concorde Building
Dinan Street, Verdun
Beirut, Lebanon
Tel: 00 961 1 752 100

contents

1

Olive Oil: Where It All Began

Olive Oil: Where It All Began

" They are twisted, they kneel to pray, and they raise their arms, members tyrannized by movement, all elbows and knees. The bent roots suck the golden oil from the heart of the earth for the lamps of the saints and the salad of the poor "

Stratis Myrivillis, Greek poet

The olive plays a fundamental role in Lebanese life, and olive oil has been treasured for generations. It is certainly valued in my family. I will never forget the story my husband Henri told me, of how his grandparents left their village of Hadath el-Jobbeh in northern Lebanon to seek a new life in Africa during the 1920s. As they sailed overseas, leaving all their possessions behind, they insisted on taking a few gallons of their precious olive oil with them. Now, whenever their grandson travels for work, he always takes some with him, too. He insists that the taste of Lebanese olive oil is unsurpassed.

Years ago, when Henri was driving down the mountain from Faraya Mzaar, Lebanon's largest ski resort, he came across an old olive wood trunk, abandoned in the fields, destined to be sold as firewood by some farmer. With help from our dear friend and restoration expert Nagi Chartouny, he turned the trunk into the base for our unique dining table.

I imagine that many of you only know Lebanon from the news headlines as a country torn by war and riddled with political strife. In fact, it is one of the most diverse and hospitable countries in the world and hides amazing riches within its borders.

Lebanon is a country of some four million people all crammed into just 10,452 km² of land along the Eastern Mediterranean between Syria and Israel. Lebanon is shaped by the twin mountain ranges that separate it from Syria. Its heavily inhabited, narrow coast is lined with orange and banana plantations, behind which the slopes of Mount Lebanon, the range that gives the country its name, rises steeply. The mountains, which are carpeted with pine trees and olive groves, rise to a maximum height of just over 3,000 metres and are covered in snow each winter. Lebanon is blessed with plentiful supplies of water and is the only country in the region that has no desert. The Bekaa Valley, which lies between Lebanon's twin mountain ranges, is one of the most fertile regions in the Middle East.

Left:
So ubiquitous is the olive tree, from ancient colossi to younger striplings, that it is impossible to imagine Lebanon's countryside without them.

Right:
Remnants of early olive presses still exist in some villages.

Because of its natural bounty, Lebanon has always attracted outsiders. It has been invaded by everyone from the Egyptians in the 13th century BC, through to the Assyrians, the Persians, the Romans, the Ottomans and finally the French, who withdrew in 1946. The mountains have traditionally provided refuge for those fleeing war or persecution, and Lebanon's villagers are every bit as welcoming today. Coastal dwellers have been traders and entrepreneurs ever since their Canaanite and Phoenician ancestors took to the seas thousands of years ago and even now, the Lebanese are renowned for their culture of enterprise and business.

Although the old ways of the mountains are fast disappearing - speeded on by the exodus of younger generations unable to resist the lure of the city or the promise of foreign countries - farmers have held on to their ancestral way of life. Many continue to herd goats and sheep, plant grain and cultivate fruits, vines and olive groves the way they have for centuries.

The countryside is full of olive trees, some centuries old, some more recently planted; an estimated 13 million olive trees account for 20% of Lebanon's total cultivated area. Almost everyone you speak to in Lebanon, from city executives to the village grocer, has a connection to olives. They either own a couple of trees themselves - or at least their neighbour does - or they own an immense grove somewhere in the mountains.

Most of the oil that comes from them is produced using traditional methods of cultivation and pressing, handed down from generation to generation, that elsewhere would be deemed outdated, although more modern methods have also been gradually introduced in recent years.

Olive oil is essential in the Lebanese kitchen. Most homes have a bowl of *zaatar* (dried wild thyme) and olive oil at hand for dipping bread and generally have several bottles stashed in a cupboard for drizzling over plates of *labneh,* a type of strained yoghurt which is the staple of any Lebanese table. All Lebanese emigrants miss their oil and there have been many attempts to market it abroad, but it is fair to say that for the most part, Lebanese olive oil is underappreciated and has yet to attract global consumers. That may soon change. Just as people are becoming more discerning with regards to Lebanese wines, Lebanese olive oil lovers are increasingly demanding only the best. The local industry's reputation has grown rapidly and as Lebanese olive oil becomes better tasting, and of higher quality, it may begin to attract the global recognition it deserves.

The mild winters and temperate summers of Lebanon's mountains, where the bulk of olives grow, are perfectly suited to growing olives. So ubiquitous is the olive tree, from ancient colossi to younger striplings, that it is impossible to imagine Lebanon's countryside without them. They are crucial to the village way of life, which is closely tied to the cycles of cultivation. Olives and olive oil are also essential to the *moune,* the winter preserves and pickles prepared by villagers each autumn.

My journey of discovery into the world of Lebanon's olive traditions began with a meeting with Dr Rami Zurayk, professor of agriculture at the American University of Beirut. An advocate of organic produce, he introduced me to a range of olive oils marketed under the name, *La Route des Oliviers,* the Olive Route. I took these oils as the starting point of a journey that I now want to share with you - a route that will allow you to discover Lebanon not as the biblical 'Land of Milk and Honey' but rather as the 'Land of the Olive'. It is a journey that took me from the beautiful hidden valleys of the South to the terraced terrains of the North, every step of the way leading me to welcoming and hard-working people who will remain etched in my memory, and hopefully in yours too, forever.

Lebanon is home to 19 different, and sometimes conflicting, religious groups but its villagers are all united in their overwhelming hospitality and love of the land. Whilst researching this book, I entered the homes of perfect strangers who were only too happy to tell me their stories over cups of Arabic coffee or tea and often a tiny glass of their olive oil as well. Such villagers have resisted the winds of change and remain true to their traditions, leading resolutely independent lives.

This book is not intended as a comprehensive reference to olives and olive oil. It should be viewed instead as an attempt to capture the flavour of Lebanon, past and present. It is an introduction to a country where every road leads to an olive tree and will hopefully spur any reader, Lebanese or foreign, to discover the natural bounty of Lebanon's countryside and especially its greatest treasure: rich, golden olive oil. Finally, this book is a tribute to the magnificent people - enthusiasts, visionaries, agricultural experts and average farmers just trying to make a living - who bring that liquid bounty to our tables year after year.

The Origin of the Olive Tree

The *Olea Europaea*, the olive tree as we know it today, is one of the oldest cultivated trees in the world. It is descended from the oleaster, which was just a humble Levantine evergreen until some clever farmer grafted it with a fruit-producing tree and created the olive.

Lebanon's soil and climate is ideally suited to olives, reasons why they flourish here like nowhere else; but even when treated badly by Nature or by Man, this tree maintains its will to survive. The olive, you see, borders on the immortal.

Ancient hollow tree trunks, filled with rocks to keep them standing, are a familiar sight on Lebanon's mountain terraces - if you take the time for a closer look, you'll see that what appeared from afar to be a dead tree, sports new twigs spouting fresh leaves.

The northern villages of Amioun and Bshaale are home to some of the oldest olive trees in the world, many of which date back at least 1,500 years. If you have visited Lebanon, you may well have driven past them without giving them a second thought, but generations of villagers have cherished these ancient survivors and have kept them standing even when they no longer produced fruit, while village children have for centuries played in their hollow trunks and hidden their treasures in them. We have become so accustomed to these natural monuments that it is hard to imagine a time when they were not here.

Prehistory and Antiquity

Bronze Age - Canaanite Period (3500 - 1200 BC)
Olive oil shipped to Egypt for trade. It is used to anoint kings, embalm the dead and light temple lamps.
A wall painting in a Theban tomb depicts olive cultivation.
Iron Age - Phoenician Period (1200 - 300 BC)
Phoenicians introduce Greece to the alphabet and the olive. They spread olive cultivation around the Mediterranean.
Hellenistic Period (300 BC - 64 BC)
An olive branch and oil is awarded to winners of the Olympic games. New pressing methods are developed in the Levant.
Roman Period (64 BC - 399 AD)
Romans develop a taste-based classification system for olive oil.
Byzantine Period (399 - 636 AD)
Thousands of distributors trade in olive oil and use in cooking, massage, body-care and lighting is widespread.

Right:
Old pressing methods using stone mills are still common place throughout Lebanon.

Far right:
Ancient hollow tree trunks are a familiar sight on Lebanon's mountain terraces, while village children have played in their hollow trunks and hidden their treasures in them.

The Ancient Roots of the Olive Tree

Standing on Lebanese soil, you get the feeling that this country is thousands of years deep and that beneath you lie centuries of hidden treasures.

One of those treasures may be the first olive trees known to man. The olive's exact origins are a matter of debate, but the general consensus is that the first trees were cultivated somewhere in the Levant thousands of years ago.

Certainly, it is true that the Phoenicians, one of the ancient inhabitants of modern Lebanon, played a major role in spreading the tree around the Mediterranean. As far back as 1600 BC, they introduced olive trees to the Greek isles and later to the Greek mainland, Italy, southern France, Spain and finally North Africa.

For many centuries, the Phoenicians were the main traders of the Mediterranean and the Middle East. They established trade colonies in Spain and North Africa and on many Mediterranean islands, including Cyprus, Sardinia and Malta. These early merchants laid the foundations of many Mediterranean port towns, including Marseille in southern France, which today is a major producer of olive oil soap.

During the Bronze Age (3500 -1200 BC) the coast of the Levant was inhabited by the Canaanites, a people thought to be descended either from Chalcolithic (4500 - 3500 BC) or Neolithic (7500 - 4500 BC) settlers.

One of their settlements, Byblos, a coastal town in central Lebanon, became the major port of the Mediterranean. From here, cedar wood and olive oil were exported to Egypt, the former for shipbuilding and tomb construction, and the latter for use in funeral rituals and cosmetics. In return, Egypt sent gold and papyrus to Lebanon.

A wall painting in a late Bronze Age tomb in Thebes, which dates to the reign of the Pharaoh Amenhotep, attests to this trade and shows Canaanite merchants carrying jars from their boat to the marketplace. Jars like those in the painting have been found during excavations in Athens and are believed to have contained olive oil. This makes sense, as the Egyptians did not produce olive oil themselves. The Egyptians used olive oil to light temple lamps, for conditioning the skin and hair and also for massage and embalming.

The Levant has been home to the olive tree for millenia. Sites in the ancient southern Lebanese city of Sidon contain evidence of clay jars, used for storing olive oil, dating back to the Bronze Age. Both the museum at the American University of Beirut and the Beirut National Museum have such jars on display, as well as smaller vessels used as oil lamps. In the village of Oumm el-Amed, there is a stone basin used for the crushing of olives that is thought to date back to the Hellenistic period (300 - 64 BC), further evidence that the Lebanese were pioneers of olive oil pressing methods. Meanwhile, at archaeological sites in Beirut, traces of olive wood and carbonized olive pips dating back as far as Roman times have been uncovered.

The most significant discoveries, however, were in the southern Lebanese villages of Shimm, Marjiyat and Ras el-Ain, where large pottery jars used for storing olive oil dating back to the 4th millennium were found. Archaeologists even unearthed the remains of Roman houses and a temple as well as several intact olive presses - the oldest ones found in Lebanon so far - and stone basins with shallow channels, all of which suggests that southern Lebanon was a significant centre of olive oil production in Antiquity. Khan Khalde, a few kilometres south of Beirut, and the Metn region mountain village of Beit Mery, which had its own Roman olive presses, also appear to have been centres of oil production.

Above: *From the port of Byblos, cedar wood and olive oil were exported to Egypt - the former for shipbuilding and tomb construction the latter for use in funeral rituals and cosmetics.*

Right: *Many trees are split in half, as if they have been hit by lightening, but this is due to old age. They stand, almost two separate beings.*

Left:
The ancient Egyptians were among the first to benefit from olive oil, which the Canaanites shipped from Lebanon in amphorae jars.

The southern Bekaa Valley village of Rashaya el-Fukhar has been producing *qahf* and *habiya* pottery jars for storing olive oil for centuries, and archaeologists have found remains in the area that date back even further. Ancient jars held up to 200 litres of olive oil and are similar to the ones in which the villagers today store their *moune*, the preserved and pickled winter supplies.

Myths, Symbols and Legends

Divinity seems to flow through the branches of the olive tree. It has been used as a religious symbol of peace, life or fertility by all of the civilizations of the Near East and Mediterranean. In ancient Greece, courageous soldiers were honoured with a crown made of an olive branch; mythology tells us that Hercules, who is credited with founding the Olympic games, gave them to valiant players as a celebration of their victory. The first Olympic torch was a burning olive branch. Later, the Romans used the olive branch as a symbol of peace and the tree was considered so sacred that those found guilty of cutting one down were condemned to death or exile. For early Christians, the olive was a symbol of renewal. It was with an olive branch delivered by a dove that God told Noah of the end of the Flood and the beginning of a new world.

Olive oil has been regarded as sacred for thousands of years. Modern excavations of Egyptian tombs have unearthed containers of olive oil. As Egyptians were only buried with their treasures, this is proof of how precious this substance was. In the temples of Baalbek in Lebanon's Bekaa Valley, oil was offered to the gods at the end of a good harvest to give thanks, and since Antiquity, olive oil lamps have been used to light temples and later, churches and mosques.

The ancient Egyptians believed that it was Isis, the Mother of the Universe, who taught mankind to extract the oil from the olives. Greek legend tells of how Athena, the goddess of Wisdom, during a dispute with Poseidon, god of the Sea, over ownership of the land where Athens was later built, planted an olive tree where the Acropolis now stands. The olive tree that grows near Athena's temple on the Acropolis, the Erechtheion, is supposedly descended from the tree that the goddess and founder of Athens planted there millennia ago.

The belief that olive oil conferred strength and youth was widespread, which is why ancient Greek athletes rubbed olive oil on their bodies before tournaments. All over the ancient world, olive oil was infused with flowers and grasses to produce medicine and cosmetics. A text found in Mycenae, in the Peloponnesus, listed aromatics like fennel, sesame, celery, watercress, mint, sage, rose and juniper as some of the ingredients added to olive oil in the preparation of ointments.

Literary references to the olive tree date at least to the Greek philosopher Sophocles (496 - 406 BC) who referred to the tree in his play *Oedipus at Colonus*. In Homer's *Iliad,* the olive gets numerous mentions, proving its importance to the Greek economy - although in my battered old translation, I have never come across the "liquid gold" reference mentioned by so many writers, but perhaps this is only in the original version.

I visited Bshaale, a small village in northern Lebanon, where local belief has it that the massive olive tree on a hill overlooking the village is the oldest in the country and possibly the world. They say this is the tree from which the dove took the branch to Noah. This seems more plausible if you believe, as many Lebanese do, that Noah eventually died in the Bekaa Valley.

When visiting another town in northern Lebanon, I was told a story by a village elder that linked the Biblical figure Adam to olive oil. Apparently, the First Man suffered from a pain and complained to God. In reply, the Angel Gabriel descended from Heaven with an olive tree. He gave it to Adam and told him to plant it, then pick the fruit and press out its oil. This oil, said Gabriel, would be the cure for his pain. Perhaps this is why so many Lebanese believe that a dose of olive oil a day keeps the doctor away.

Right:
Sites in the ancient Lebanese city of Byblos contain evidence of ancient olive oil presses.

The Bible itself is packed with over 140 references to the value of olives and olive oil. In the Old Testament, it is said that men who worked the olive harvest were exempt from the army. In Jeremiah, the olive becomes a metaphor:

"The Lord called thy name, a green olive tree, fair, and of goodly fruit: with the noise of a great tumult he hath kindled fire upon it, and the branches of it are broken."(11:16)

Elsewhere, it is a source of sorrow, as we learn that the worshippers of the pagan god Baal are punished by having their most precious possession, their olive trees, taken from them; it was on the Mount of Olives, just outside the walls of Jerusalem, that Jesus spent that fateful evening with his apostles, waiting for Judas and the Roman soldiers. Trees, which are believed to date from that time, some 2,000 years ago, still stand today in the garden of the church built to mark the site.

Even today, the olive plays an important role in Eastern Christianity. During Lebanese Easter celebrations, olive branches are handed out on the last Sunday of Lent in memory of the arrival of Jesus in Jerusalem. Oil is blessed by bishops and then distributed to churches to be used during baptisms and confirmations, to ordain priests and consecrate churches, as well as for anointing the sick and the dying. Pilgrims visiting religious sites like Harissa and Mar Charbel in northern Lebanon take home small bags containing oil-soaked cotton balls, which they give to other family members as a blessing. The bag is often put under a child's pillow to ward off evil spirits.

During a Greek Orthodox wedding, the couple are anointed with oil, much in the way newborn babies are baptised in a Maronite (Lebanese Catholic) or Roman Catholic ceremony. In both cases, this signifies entry to a new life. For their part, Greek Orthodox babies get a more comprehensive soaking and are completely covered in oil. If a part of the body is missed, it is said that area will remain weak for the rest of the child's life. The parallels with the ancient Greek myth of Achilles and his weak heel, the only part of his body his mother had forgotten to dip in the death-defying waters of the river Styx, are unmistakable.

It isn't just Christians who value the olive. In addition to being named as one of the trees growing in Paradise, both the tree and the oil are often mentioned in the Koran. In the Sura of the Light (24:35), olive oil appears as a symbol of light:

"God is the light of the heavens and earth ... lit from a blessed tree, an olive, neither of the East or West, whose oil is well-nigh luminous, though fire scarce touched it ..." Oil fuels the lamp, which gives light. As light is divine, so is the oil that lights the lamp.

The olive was also the blessed tree of the prophet Mohammed.

"Olives and pomegranates, similar (in kind) and different (in variety): eat of their fruit in their season, but render the dues that are proper on the day that the harvest is gathered. But waste not by excess: for God loveth not the wasters." *Sura of the Cattle* (6:141).

Today, olive oil is given as a traditional gift to pilgrims coming to Mecca for the Hajj.

Used in Lebanon in various rites of passage, olive oil plays a role in the rituals surrounding birth and marriage. For example, Lebanese poet Kahlil Gibran pays tribute to it in "*You have your Lebanon and I have my Lebanon*," where he refers to the olive's eternal qualities, which outlast mankind. Even the legendary Lebanese singer Fairouz pays tribute to this tree in a song on her album "*Mish Kayan Hayek T'Koun.*"

If its spiritual significance is shared by many religions in the Middle East, both past and present, the power of the olive branch as a Roman symbol of peace and hope has transcended its early origins to become universal. Today, all over the world people talk about "offering an olive branch" when they wish to make peace. In Lebanon, olive groves were planted right up to the barbed wire border with Israel, after the latter withdrew from southern Lebanon in 2000 after an 18-year occupation. Liberation, rebirth, renewal, peace: the olive tree encompasses them all.

2
Living Legend: *min sajra lal hajar* from tree to stone

Living Legend: *min sajra lal hajar* - from tree to stone

One of the finest times to be in Lebanon is October. There is a saying that between October and November there is a second summer: *bein teshreen wa teshreen seif tenni*. The oppressive heat has gone and the first rains arrive, announcing the impending olive harvest for many of the farmers. In the mountains, the countryside is untouched by modernity and olive trees grow in the wild as they have done for millennia. They can withstand scorching summers and harsh winters, but ideally favour a warm dry climate with seasonal rainfall. Which is why the tree flourishes in the Mediterranean region, where they grow along the coast and at altitudes of up to 1,000 metres.

This tree almost has a mind of its own. It refuses to comply with the rules of nature. On one farmer's land it will not prosper, but just a few hundred metres away, groves are lush and thriving. Some years, a tree may be weighed down with fruit, while in others it bears but a handful. Farmers indulge their trees, treating them with the patience reserved for children. They tend them as best as they are able and each has their own idea about how many times a year to plough the fields, how best to prune the trees or the best time to harvest.

Olive and olive oil production levels vary from year to year in all Mediterranean countries. The tendency is to alternate years of heavy and light harvests. Such fluctuations in Lebanon are higher than elsewhere for two reasons: Lebanese olive groves are mostly rain fed (if they were irrigated there would be less fluctuation) and the pruning and harvesting methods tend to be traditional and somewhat unsophisticated.

Statistics for Lebanese olive oil production are also somewhat 'variable.' To get a better picture, I decide to head north to the town of Zgharta to meet Sheikh Sleiman El-Dagher, president of SILO (Syndicate of Interprofessional Lebanese Olive Oil Producers).

With map in hand, I tell my driver for the day, Mohammed, to take the road from Tripoli. Of course, he knows a faster route via Koura. So we take the turn after Shekka and head up into the hills. The route is long and winding, with plunging forested valleys to the right and immense olive groves all around. I settle back to enjoy the view until I'm jolted out of my reverie by the sight of the snow-covered mountains behind al-Arz, Lebanon's famed Cedars, and I realize that we have travelled too far up into the mountains.

"Maaleshi," says Mohammed. Never mind. He turns back and asks for directions at a Lebanese army checkpoint. When we arrive in the village of Kfarzina, he asks a middle-aged couple sitting by the side of the road selling *foul* (broad beans) and strawberries from the back of a beaten up pickup truck, for the house of the Sheikh (any prominent person in Lebanon is respectfully referred to as 'Sheikh'). They know him - all the mountain people know their local olive growers.

We pass the village square where a statue of St George stands prominently in front of the church. Adonis, the name the Greeks gave the Semitic god Tammuz (who was transformed into St George by the Christians, and later into *El Khudr*, the Green One, by the Muslims), used to hunt in these mountains. For thousands of years, people would pray to Adonis for health and fertility. Now they make that same request to St George, which is perhaps why all the villages in this part of the world seem to have a church named after the saint.

We find the way to Sheikh Sleiman's sizeable house, which is perched on top of a hill overlooking endless groves of olives. It's a breathtaking location. El-Dagher's life revolves around olives, but he has taken the morning off to stay at home to meet me. This is typical of the Lebanese mountain people, who are so generous with their time.

Left:
Leaves and twigs are removed from the olives before they are placed in a crate.

Above:
One of Lebanon's oldest olive trees stands below the Mar Youhanna (St John) church set on a rocky cliff studded with Roman tombs.

"Bonjour," his French wife greets me as she takes me into the house to meet her husband. I attempt to speak to her in her native tongue, as she keeps me supplied with Arabic coffee and sweets, but I keep lapsing back into Arabic. The welcome formalities over, I start my questioning and soon discover that El-Dagher is well informed, with incredible insight into the Lebanese olive oil industry.

He owns many olive groves, produces his own oil, and can quote facts and figures effortlessly. He estimates that around 15% of all olive orchards are less than 10 years old, that 60% are between 10 and 50 years old and the remainder are anything from 50 to over a 100 years old. There are trees in Lebanon that are 150 years old and still producing fine crops of olives - although in general, fruit production decreases once the tree reaches 70 years of age.

El-Dagher tells me that in recent years, many new trees have been planted in southern Lebanon, in the Akkar in the North and in the Bekaa Valley through government-funded expansion plans. Significant increases in oil production are expected in these area once the trees mature in five years or so.

According to El-Dagher, olive production in Lebanon rose from 30,000 metric tons in 1998 to a high of 190,000 metric tons in 2000. In 2002, total olive oil production amounted to 25,816 metric tons but the following year, this figure had dropped to about 7,000 metric tons due to a poor harvest. In 2004, olive oil production rebounded and reached about 26,000 metric tons. Of that total, Extra Virgin olive oil accounted for about 2,750 metric tons; it generally accounts for 10-15% of all olive oil produced in Lebanon.

Due to its popularity, most Lebanese olive oil is consumed locally. Yearly consumption was calculated in 2005 at about five litres per head. National annual consumption is about 15,000 tons.

"Back in 1960 though, the yearly consumption per head was about 13 litres in Lebanon," says El-Dagher. The decrease has been due to rising popularity of sunflower, corn and soya oils, but El-Dagher forecasts an increase to at least six litres per head by 2010 as the Lebanese realise the health benefits of olive oil.

He emphasizes the fluctuations, lamenting that in 2005, many areas experienced a harvest that was about only 5% of normal. He goes on to say that olive oil production is mostly a family business and

Above:
Sheltering under the boughs of an olive tree in the Koura, two friends, Daliah and Elle, are bound by friendship and their love of nature.

its seasonal nature means that the smaller-scale producers need to have a second job to be able to support themselves. Gesturing at the surrounding groves, which remind me of Tuscany and stretch far below where we stand, he continues.

"This is the major olive growing area in Lebanon and in order for our farmers to survive, local production has to be protected from cheap foreign imports."

As I leave, he offers me an orange from a tree in his garden. I see a wooden rosary hanging on a branch.

"Is that for protection?" I ask him, hoping to hear some ancient myth about the saints, the evil eye and the power of the rosary to ensure a good harvest.

"No," he replies, "perhaps one of the children put it there when they were playing in the garden."

I try, I suspect unsuccessfully, to hide my disappointment.

Local Varieties

An olive tree needs to be eight years old before it bears fruit that can produce oil. The varietals planted in Lebanon generally yield an oil content of around 20%. Unlike elsewhere, local varieties of olives are not differentiated by area or region, and several different types are often grown side by side in the same grove. Despite looking alike, different varieties of olive produce fruit that is different in appearance and taste. The most common variety in Lebanon is the Souri, which bears fruit with a 5-30% oil content and produces oil with a distinctive flavour. About 70% of all olive trees in the country are Souri. Its name is derived from the Arabic name for Sidon, a port town in southern Lebanon.

The varieties grown in Lebanon include:

- Baladi (20% oil content). The name means "local" in Arabic. It produces oil with an herbal aroma.
- Shami (under 15% oil content). It produces a sweeter oil with a more subtle flavour.
- Ayrouni (15% oil content) and Smoukmouki (under 15% oil content). Both produce oil with a pungent taste.

Additionally, some Shetawi olive trees are cultivated in Lebanon. Some farmers graft Baladi branches onto Shetawi trunks to produce larger and fuller fruits.

There are about 345 varieties of olive trees in the world. Like wine, olive oils come in vintages and differ from year to year, although certain characteristics in the taste never change. Unlike wine though, olive oil does not improve with age and should not be kept for more than two years. It is best consumed young and fresh, when no oxidation has taken place. Reputable suppliers will stamp a production and expiry date on the bottle to help customers buy the freshest oils.

So why the difference in flavours? Is it only the variety of olive that determines taste? What of soil, climate, or even extraction techniques?

I repeat some of the theories to agricultural engineer and manager of Olive Trade, Youssef Fares. He grows olives in the Akkar region of northern Lebanon and is the owner of a modern oil press that he uses to extract his own brand of olive oil, Zejd. His organic olive oil is rated highly and it is on the shelves of specialty stores worldwide.

I tell him that all the farmers I've met claim that their oil is the best, and that when I've asked them what makes their oil so special, they tell me things like, "it's the sun in our area," "it's the height of our groves," "it's our traditional ploughing methods" or "it's the flowers, plants and vegetables that grow nearby."

Some olive oil experts have started using the term 'terroir', a French word normally associated with wine, to categorise different olive oils. They believe that oil from the same variety will taste different if the tree is planted in moist red soil or in drier calcareous soil. Others believe that the taste is mainly influenced by the variety cultivated and the techniques used to extract the oil.

"To a certain extent, high-quality olive oil is considered a 'terroir,' product like wine. Terroir refers to a select, genuine, organic and natural product produced in limited quantities, cultivated in a traditional fashion, within a specific geographic region, with a rich historic background related to its origins," explains Fares. "But of course each farmer will have their own ideas about what makes their oil the best. Ultimately, it is their *savoir faire* that will determine the quality of their oil."

Left:
A handful of olives is as precious as a handful of gold.

Right, clockwise from bottom left :
The blossoming season lasts for about two to three weeks, during which the branches are covered in bunches of lightly-scented ivory pearls which burst into flower.

In Lebanon, local varieties of olives are not differentiated by area or region, and several types are often grown side by side in the same grove.

Gathered together after the harvest, the olives are now ready to go to the press.

The Yearly Cycle

The olive season in Lebanon begins with the yearly pruning, which usually takes place in February or March. One warm spring afternoon, I watched a farmer prune his trees and throw the trimmings into a pile. He told me that he uses the off-cuts for fires on cold nights, as olive wood gives off a good deal of heat. In fact, it is so good for making fires that other farmers slow-burn the trimmings for several days in an earthen oven and make charcoal for barbeques and tobacco waterpipes. He then told me he uses the ashes from his fires as fertilizer for his olive trees.

"All the parts of the olive are useful," he added, "just like a goat."

If a tree is left in its wild state, it will still produce olives but the fruit will be small. When a tree is producing fruit, it has less energy for new branches. In order to grow, wild trees normally produce fruits on alternate years, one year for growth, one year for olives. Proper pruning produces larger olives and a more even distribution of fruit by letting the sun into the foliage, and reduces the height of the tree, which makes picking easier. It also encourages trees to fruit every year. In Lebanon, improper pruning means that many trees do not produce as much as they could.

Left:
The most common olive variety in Lebanon is the Souri; its name is derived from the Arabic name of Sidon, a port town in southern Lebanon.

Right:
Caring patiently for his land, the Sheikh is continuing the work of his forefathers.

Back in Beirut, I have a meeting with Hussein Hoteit, an agriculture engineer who consults on various rural development projects and lends his expertise to non-governmental organisations (NGOs), amongst them the RMF (René Moawad Foundation), Mercy Corps, the ICU (Institute for University Cooperation) and SRI International (Stanford Research Institute). He also does consultancy work for the Lebanese Ministry of Agriculture.

Hoteit tells me that he is distributing a leaflet to farmers about new pruning methods and that he is also trying to encourage the use of a portable pneumatic device that can be used to prune trees and to harvest olives. It resembles a leaf-blower and comes with a shearing attachment that can be used to trim branches. When the attachment is removed, pressurised air is used to blow olives off the trees.

Technological innovations aside, there are still many farmers who barely bother to prune at all.

"They might just pick a few twigs off here and there after the harvest, to tidy up," he explains, "but this is not sufficient if they want to improve yields."

Mature olive trees have roots strong enough to survive without irrigation. Seasonal rains, air humidity and moisture from melting frost generally provide enough water. Irrigation would ensure annual crops and larger loads, but most Lebanese farmers are not able to incur the costs of installing a system and so prefer to rely on yearly rainfall, which is high during the winter months.

The higher up olives are planted, the less exposed they are to parasites. The Ayrouni, Baladi and Souri varietals are especially resilient to the colder air of the mountains. Trees planted on the coast or at low attitudes tend to need more attention. Some farmers treat their trees each February with chemical fertilizers, while others follow organic methods.

Hoteit is an advocate of the latter. He tries to get farmers to use a system of 'green' manure. Leguminous seeds are planted under the trees in November and as the plants sprout in the spring, they are ploughed up and reincorporated into the soil as a natural fertilizer, but not everything can be dealt with in an environmentally-friendly way.

"The Olive Fly is the biggest threat, and once it attacks the tree the only way to kill them is with pesticide," explains Hoteit.

Walid Nassour, an assistant at the World Vision Lebanon Centre (an NGO set up to improve organic farming conditions) in Marjayoun in southern Lebanon explained to me that Olive Fly traps, also known as McPhail Traps, are put out in July. During an expedition to Deir Mimas in the Marjayoun district, I was able to observe how this is done.

The traps contain liquid pheromones that attract the flies. They stick to the liquid and are caught in the trap. Many farmers use empty plastic bottles in place of McPhail Traps, to keep costs down, and you will often see old water bottles dangling from olive trees.

After their long winter hibernation, olive trees bloom in April. The branches are covered in tiny lightly-scented ivory pearls in bunches of 10 or 20, which burst into flower - each bunch of blossoms produces only one or two olives. The blossoming season lasts for about two to three weeks, and during this time the pollen drifts between the trees on the breeze. As the leaves turn from silver to light green, the tree begins the long process of producing its crop of olives.

Left:
Once the olives have been picked, the leaves and twigs are removed before the crop is taken to the local press.

Harvest Time: *Az-zeitoun*

As the summer fades and autumn sets in, farmers pay close attention to the state of the ripening olives, waiting patiently for the right day to begin the harvest.

A French peasant proverb says that on St Catherine's Day (the 25th of November), "the oil is in the olive," meaning that this is the best day to harvest. A Lebanese proverb says that the best time to harvest is on the 1st of November, All Saints' Day. Most farmers laugh off these axioms and harvest after the first rain, as the olives are then easier to pick. All farmers agree on one point though: because olives ripen at different times, the correct time to harvest varies.

Each variety has different harvest dates. The Cailletier Olive, which is grown in the south of France, is harvested in January as it takes longer to mature. The Souri, so popular in Lebanon, is best harvested between October and November.

Olives picked early produce an oil that is stronger in taste and greener in colour. It contains more phenols, a natural preservative, and will be less acid and thus has excellent health benefits. However, olives picked at an early date also contain less oil, so some farmers prefer to pick later to get more from their trees.

In recent years, members of local village cooperatives in Lebanon have begun to attend lectures held to increase awareness of the benefits of harvesting early to produce Extra Virgin oil, which fetches a higher market price. The timing of the harvest and the method of picking is of utmost importance. The best time is late autumn. Farmers are advised to check the olives daily and to harvest as soon as the skin darkens. At this time, the fruit contains the most oil and the antioxidant polyphenols. This is when it will be of the best quality. The olives should always be picked directly from the tree. Those fallen to the ground will not produce Extra Virgin oil.

The olive harvest has always been central to the lives of villagers. Some men work as fishermen during the summer and then return home to manage their trees. In Lebanon, traditional harvest methods are still commonplace, unlike in other countries, where the harvest has mostly been mechanised. So the picking schedule is determined by the availability of Lebanon's migrant workers, who are only available after the grape harvest in September. Relatives, friends, hired labour and even village children - who are glad for any excuse to skip school - also help out, joining together in this celebration of natural bounty. Already promoted as an eco-tourism event all over Europe, the olive harvest is now on the calendar of Lebanese rural tourism activities.

I once heard a saying: "Your wife, your dog and the olive tree ... the more you beat them, the better they become." None of the farmers I repeated this to had ever heard this before, but they all laughed when I told them.

It goes without saying that wife beating is totally unacceptable in Lebanon, and now even the beating of olive trees (the easiest way to make the trees shake off their fruit) is slowly becoming taboo, as farmers realize that the tree's tissue and the olives themselves can be damaged in the process. Tree beating can also lessen the following year's harvest. It is still widespread though. I observed the process in a grove in Hasbaya, a town in the south-western corner of Lebanon, where a farmer told me that beating the fruit off the tree with a stick is faster and reduces the cost of hired manual labour, an expense many farmers can ill afford. Those who can afford modern equipment use mechanical shakers, but machine harvesters are effective only in groves wide enough to allow them to be driven through. Such groves are uncommon in Lebanon. Some farmers do however use the pneumatic olive blowers Hussein Hoteit has been trying to encourage, but the preferred method for harvesting in Lebanon remains the age-old tradition of picking by hand.

The ideal time to start picking is early in the morning - once the olives have soaked up the heat of the day, they decay more quickly. Workers arrive at the groves at dawn. Some come in overloaded *service* (shared) taxis or in battered VW Campervans. Others come on foot. Women arrive in groups carrying baskets, with their provisions for the day: bread and cheese, as they know they will not leave the groves before sunset.

Left:
The preferred method for harvesting in Lebanon remains the age-old tradition of picking by hand.

Right:
At the press, the olives are washed and any remaining debris or twigs are removed.

Olive picking requires care and is time consuming. The women begin their day by picking up the olives that have fallen to the ground. Olives that fall to the ground, or onto the nets or canvas spread out beneath the trees to catch falling fruit, should be packed up as soon as possible, as oxidation can occur if they are left on the ground. The men climb the trees with the help of ladders to reach olives that are out of reach. They are stroked off the branches with a plastic comb or gently pulled off by hand. Harvesting a tree can take between 30 minutes to four hours depending on its size, width and height. Once picked, the olives are placed straight into a basket. In the past, they used to be put in a sack made out of goat leather, called the *jrab*. This has now been replaced by straw or wicker baskets. When olives are placed in nylon sacks, they tend to rot if they are not pressed within a couple of hours and the quality of the oil will suffer as a result. Farmers increasingly use plastic crates as they accumulate less moisture than wood and thus keep the olives drier and less likely to rot.

Hussein Hoteit, who has groves of his own, explains how even plastic crates can be used in an environmental way.

"I borrow crates from a friend who has apple orchards," he says, "and in return I give him some oil."

Leaves and twigs are removed from the olives before they are placed in the crate. While this hard work is going on, the elderly women and children are kept busy clearing the ground, which becomes littered with falling leaves and twigs.

Olives must go to the press within twelve hours. If they are not kept cool, the taste of the oil can be affected. This is easier said than done in hot, dusty Lebanon though, especially as the masses of olives being picked cause bottlenecks at the presses.

There is a tradition in the southern village of Bint Jbeil, known as *bawajeij* - children go to the olive groves after school and gather the leftover fruits to sell or barter for sweets. Many farmers, after harvesting their groves, also allow the poor local families to take the olives that are left on the tree. This practice is common all over Lebanon.

Local farmers say that you can never have a 100% harvest. A farmer once told me that either the fruit is too high up to reach, or it is just not worth employing the extra labour to do the picking.

The harvesters' hard work is usually rewarded with a feast. Never miss an invitation to lend a hand at a friend's grove, as you are sure to be treated to a delicious meal - and after, if you are lucky, a first tasting of the freshly-pressed oil. Unfortunately, in the mountains, the post-harvest feast is not as common as it used to be and harvesting has become just another task for migrant workers struggling to earn a meagre living.

GENERAL SY
TORO

The Race to Press the Traditional Way

The local olive press has always been part of rural community life. While waiting their turn, farmers chat and swap ideas, shouting to make themselves heard above the racket of the press.

There is an old Lebanese proverb that says, "if olives are not pressed the same day they are picked, they are like a bride and groom who do not make love on their wedding night."

Any farmer will tell you that if you just drop your crates of olives off at some shiny new set-up, it robs you of the full experience of the pressing. A hands-on approach does not necessarily produce a better-tasting oil but it ensures oil extraction remains a labour of love and it gives you an excuse to chat and drink coffee with your neighbours. Modern technology may be more efficient, but it is rarely the most human alternative.

Lebanon has roughly 435 traditional olive presses and 67 modern ones. Most oil presses in Lebanon still manage without machinery, and often the only concession to modernity has been to replace the donkey that previously powered the press with a motor.

Above left:
The olives are crushed, meat and stone, into a paste between two heavy stone wheels and a stone base plate.

Below left:
The paste is spread on flat woven mats, which are layered together, placed under a screw press and squeezed until a liquid substance - a combination of olive oil and water - trickles out.

Modern presses are more costly to build and to operate, but they can process three times as many olives as traditional mills. They meet international standards of production and guarantee a superior quality of oil, making them more suited to farmers aiming at the export market. That said, the top oils of the world (and the most expensive due to their increased processing time) are still pressed the traditional way.

At the press, the olives are washed and any remaining leaves or twigs are removed. Then they are crushed, meat and stone, into a paste between two heavy stone wheels and a stone base plate. This is the way it has been done for centuries.

The paste is spread on flat woven mats, which are layered together, placed under a screw press and squeezed until a liquid substance - a combination of olive oil and water - trickles out.

This liquid is put in a centrifuge to separate the water from oil. At this stage, the oil is decanted into barrels and the strong distinctive fragrance of the oil, which is different from region to region, fills the air. This is known as the 'first pressed' oil.

The remaining paste is then milled for a second time. In the past, presses were less efficient than those used today and after the first pressing, hot water was added to the paste to continue the process of extraction. This is why only the first pressing was known as 'cold pressed'. Today's technology has dispensed with first and second pressing, cold pressing and hot-water pressing, because all the oil is extracted in one go.

Zahia Abboud's press, in the ancient southern Lebanese town of Sidon, is definitely not one of those shiny new presses. A third-generation olive oil producer, Abboud's press was set up in 1932. Housed in a traditional Lebanese stone building, it conforms exactly to my idealised image of what an olive oil press ought to look like.

Straw mats lie under arches and clay jars are lined up against a stone wall, basking in the rays of the sun, waiting to be filled to the brim with golden oil.

In spring Lebanon's olive groves are carpeted with tiny pink flax, crown daisies, scarlet anemones, corn poppies and cyclamen.

"The ancestral way of pressing olives is laborious and does not necessarily produce a better-quality oil than the modern mills," admits Zahia before adding that there are many traditionalists in Lebanon who prefer to produce their oil the way their forefathers did, most of them city dwellers who only visit their village homes at the weekend.

Pressing the Modern Way

The first step of the process is still carried out the same way as it has been for centuries: the olives are separated from any leaves or other debris. They are then washed and prepared for the second stage of the process: the crushing. This is where the similarities end. In modern mills, the olives are crushed using a hammer and disk crusher that turns them into a paste. Water is added to the paste, which is churned for 45 minutes. This paste is passed into a decanter where the oil is separated from the water and solid content.

The leftover is known as 'husk'. This is used to make lamp oil and soap. Once the last drops of oil have been extracted, the dry, crumbly mixture that is left is sold as fuel.

Another extraction method, invented in the 1970s, is known as the Sinolea Process. This technique depends on fine, spinning blades. The olives are run through an electric crusher at room temperature and a metal comb is passed through the paste. The oil is collected on the spikes of the comb. The whole process takes only 20 minutes.

Whatever method of extraction they may choose, all oil producers know that the cold is their enemy. Ideally, olive oil should not be exposed to temperatures below zero, otherwise it may crystallize. If this occurs during processing, the oil will have to be re-filtered.

The Final Stage

The first dribble of oil that flows off the press into the deep marble collecting basins is awaited with great anticipation. Every farmer hopes for a first-rate oil, but it is only now, with this first taste, that they will know if their prayers have been answered. Experienced workers tending the press can recognize the quality of an oil by its smell and colour. Even before tasting, they know if it is *helou* (sweet) or pungent.

Every grower hopes that this year, their oil yield will be high. On average, an olive tree gives 20 kilos of olives and on average, 5 kilos of olives produces 1 kilo of oil. But here too, yields vary and many factors influence the amount of oil a tree produces, from the variety and the age of the tree to the weather that year and the type of press used.

Once it has been pressed, the oil can be stored in anything. Containers made of plastic, tin, glass and even clay are brought to carry the oil away. Some growers send their oil directly to large corporations, which then bottle it their own way for sale.

Like Lebanon's oils, bottle designs are also becoming increasingly sophisticated; while packaging is a matter of taste, experts advise that olive oil bottles should be dark green in colour to protect the contents from light, the main enemy of bottled oil.

Labels too are evolving. At the moment, labels in Lebanon generally list the name of the brand and the type of oil, but no further information. The vintage and a "use by" date should be displayed clearly, as oil only has a shelf life of 2 years.

Domestically, many Lebanese still buy their oil directly by the gallon from a relative or friend who owns a grove and they find labels unnecessary because they trust the producer. Traditions are hard to change and if this results in the consumption of copious amounts of oil with profits going directly to the farmer, then perhaps tradition is not such a bad thing. That said, more informative labels are vital if Lebanon is to break into the export market.

Pickled Olives

Above left: *Organic olive oil is in demand and production is on the increase.*

Below left: *The first dribble of oil that flows off the press is awaited with great anticipation by the farmer. Every farmer hopes for a first-rate oil.*

Below right: *Olive juice oozes out of the compressed mats.*

Green olives, picked during the first phase of the harvest, and black olives, gathered later on, that are not destined for pressing are sorted according to size and perfection and then pickled in brine. These olives are ready to eat after nine months to a year and can be kept for up to two years or more.

Opinion differs amongst village folk with regards to the best method of pickling olives. Some soak them overnight in a tub of water, others do not bother. Some wash the olives first and then hit them with a stone to split them before pickling, others split them first and then wash them. One villager may cut slits in the fruit with a sharp knife and then add lemon juice, lemon peel and chilli peppers to the brine. Another may recommend adding bitter orange leaves, whilst a neighbour may add garlic, bay leaves or *zaatar* (wild thyme) for flavouring. Perhaps the best method is to try them all. (See page 152.)

3
Classification and Tasting of Lebanese Olive Oil

Classification and Tasting of Lebanese Olive Oil

"Everyone will tell you that their oil is the best"

Lebanese proverb

I remember my first olive oil tasting session a few years ago. My friend Joanne took me to Ramzi Ghosn's vineyard in Tanail in the Bekaa Valley. He sells olive oil from the northern Akkar region under the Nay label. Ghosn guided us through the process, encouraging us to pay attention to the different flavours and aromas of the oils.

"Some are peppery, others are sweet," he said. "Some taste of green tomatoes and others of apples."

In fact, Lebanese oils come in many different flavours. The most common adjectives used to describe the different taste notes include honey, almond, tomato, herb, lemon, even grass and hay. Sometimes one note dominates, but in others there are so many different flavours in one oil that it is difficult to name all of them.

More pungent oils are usually extracted from olives that have been harvested early. The strong taste is the result of higher levels of polyphenols in the oil that help conservation and have significant health benefits, which is why Extra Virgin oil is deemed so beneficial. On the contrary, sweeter oils are the result of later harvesting and the pressing of more mature fruits.

IG (Indications Géographiques), a Swiss-funded program that is being implemented in partnership with the Lebanese Ministry of Economy and Trade, is exploring the special relationship that exists between an oil and its 'terroir.' Hasbaya's olive oil has been chosen as one of its pilot projects. The Ministry of Economy and Trade is seeking to award the first "appellation d'origine controlée" (literally meaning controlled place of origin; this classification guarantees the place of origin and the quality standards traditionally associated with the oil from that place) to Hasbaya's Extra Virgin olive oil.

The IOOC (International Olive Oil Council), the international organisation responsible for the worldwide marketing of olive oil, has set certain standards for defining the various grades of oil. These standards have been adopted by the Lebanese olive oil industry in order to open its oil up for export.

In general, the rules set by the IOOC for determining an oil's 'virginity' dictate that such oil should be extracted only from quality olives and that the extraction process should not make use of chemicals or involve excessive heat.

All IOOC standard oils for the export market must also pass additional taste requirements to determine their quality, and be tested for their taste and aroma, and their oleic acid content. Oleic acid is a monounsaturated fatty acid and accounts for 80% of the acids in olive oil. A low level of acidity in olive oil is considered beneficial to the health and consumption may reduce cholesterol levels and reduce the risk of heart disease. The lower the acidity, the higher the quality of the oil. Gourmets will not touch oil that contains more than 1% acidity.

Professional tasters look for the following taste defects, all of which lessen an oil's quality:

Rancidity	caused by oxidation
Mustiness	from humid storage
Metallic taste	caused by contact with metal during the extraction
Burnt flavour	caused by excessive heating
Earthy taste	from unwashed olives
Woodiness	from olives that became too dry prior to pressing
Vinegary or winey flavour	the result of olives being left too long before being pressed

Qualities and flavours tasters look for and which enhance an oil:

Pepperiness/slight bitterness - characteristic of olives pressed young, typical to Extra Virgin oil and the Baladi varietals cultivated in the Hasbaya region of southern Lebanon.

Pungency - characteristic of olives picked young wherever they are grown. Causes a biting sensation in mouth and throat when consumed.

Green banana/artichoke flavour - typical of the oil from southern Lebanon.

Green tomato flavour - characteristic of oil from the Akkar region of northern Lebanon.

Grassiness - typical of olive oil from the north of Lebanon.

Apple flavour - characteristic of oil from north Lebanon, especially the Zgharta region.

Green Almond flavour - characteristic of sweet oils. Both southern and northern Lebanon produce oil with this flavour.

Different types of olive oil

Green-coloured olive oil is produced from olives that have not fully ripened and has a strong taste. Gold-coloured olive oil is produce from fully ripened olives and has a more subtle, delicate flavour. Colour does not necessarily reflect the quality of the oil.

Extra Virgin oil - oleic acidity content of less than 0.8%. Only cold pressed. Olive oil in its purest form, with no additives.

Refined Virgin oil - cold pressed, with Oleic acidity content of between 0.8 and 2%.

Pure olive oil - a mixture of refined and Extra Virgin oil. Oleic acidity content of less than 1.5%.

Lampante oil - oleic acidity content of 3.3% or above. When chemically treated to reduce acidity, it loses its aroma. Sometimes sold on the market as Light Olive Oil. In Lebanon, Lampante Oil is used for soap making.

So Which Oil is Best?

Of course, every olive grower will insist that their oil is the best. Often emphatically. However, it is ultimately the palate that decides which olive oil is best and an excellent way to find out is to try different brands for yourself.

Most Lebanese are not aware of the difference between the various oils on the market, because they usually buy their oil from the same supplier. This is a pity, as each region in Lebanon produces its own distinctive oil.

A good way to find out what you like for yourself is to take a trip to one of the villages famous for its olives, where producers are only too happy to let you taste their oil. A good time to visit is during the pressing season. Tasting oil fresh off the press can be quite an experience: once, the oil I was given was so pungent that it triggered a coughing fit, which I tried hard to subdue so as not to offend the proud producer.

In ancient Rome, an expert taster could tell blindfolded whether an olive had been picked bare-handed or by someone wearing gloves. For those of us who are not as talented, there are several ways to make sure you experience the full flavour of an oil.

Hussein Hoteit is a qualified oil taster and does not recommend a method many farmers swear by: pouring a few drops into the palm of your hand to warm it and help release its fragrance. He says as the skin has an odour, this will affect the fragrance and taste of the oil when tasting.

"Just pour out a little (oil) into a glass," he recommends, "but not into a spoon, as the taste can be metallic."

Youssef Fares has trained with the IOCC to be a certified taster. He suggests warming the glass in the palm of your hands before tasting.

The best time for a tasting is before lunch, when your taste buds will be more sensitive. The ideal temperature oil should be at for a tasting is around 28°C. Whether you use a warm glass or cold, both Hoteit and Fares agree that you should breathe in the fragrance as you roll the oil around in your mouth, in much the same way you would with wine.

It's important to keep your mouth closed while doing this, so that none of the flavour escapes. You might pick up a hint of honey on the tip of your tongue or a subtle tomato flavour as the oil glides down your throat. Either way, your mouth will be full of the oil's aroma. If you are tasting several different oils, rinse your mouth with water between each tasting or chew a piece of apple to cleanse the palate.

Lebanon's First National Olive Oil Contest

In an effort to encourage greater domestic consumption of Lebanese Extra Virgin oils and to increase oil exports, SRI International organized the first National Extra Virgin Olive Oil Tasting Contest at the HORECA hospitality trade fair in Beirut in May 2006.

Thirty-six local producers entered the competition. Their oils were tested in laboratories to ensure they could be considered Extra Virgin. Twenty-five entrants fulfilled the criteria. The event was presided over by eight judges, ranging from agriculture engineers to professional tasters. Visitors to HORECA were also encouraged to participate in the vote.

Gaby Beainy of Jezzine in South Lebanon, who doesn't even produce oil commercially, won the prize for best-tasting oil. The winner of best presentation for bottle and label was Zeytouna (a product of the René Moawad Foundation) with Zejd coming second and the Al Wadi Gourmet and Atayeb el-Rif sharing the 3rd prize.

Below left:
Winners of HORECA 2006 olive oil contest.

Above right:
Abboud takes the first sip of the new season's olive oil.

Below right:
As bottle designs become more sophisticated, suppliers are putting the date of production and place of origin of the olive oil on the label.

4
Lebanon's Oil Producing Regions

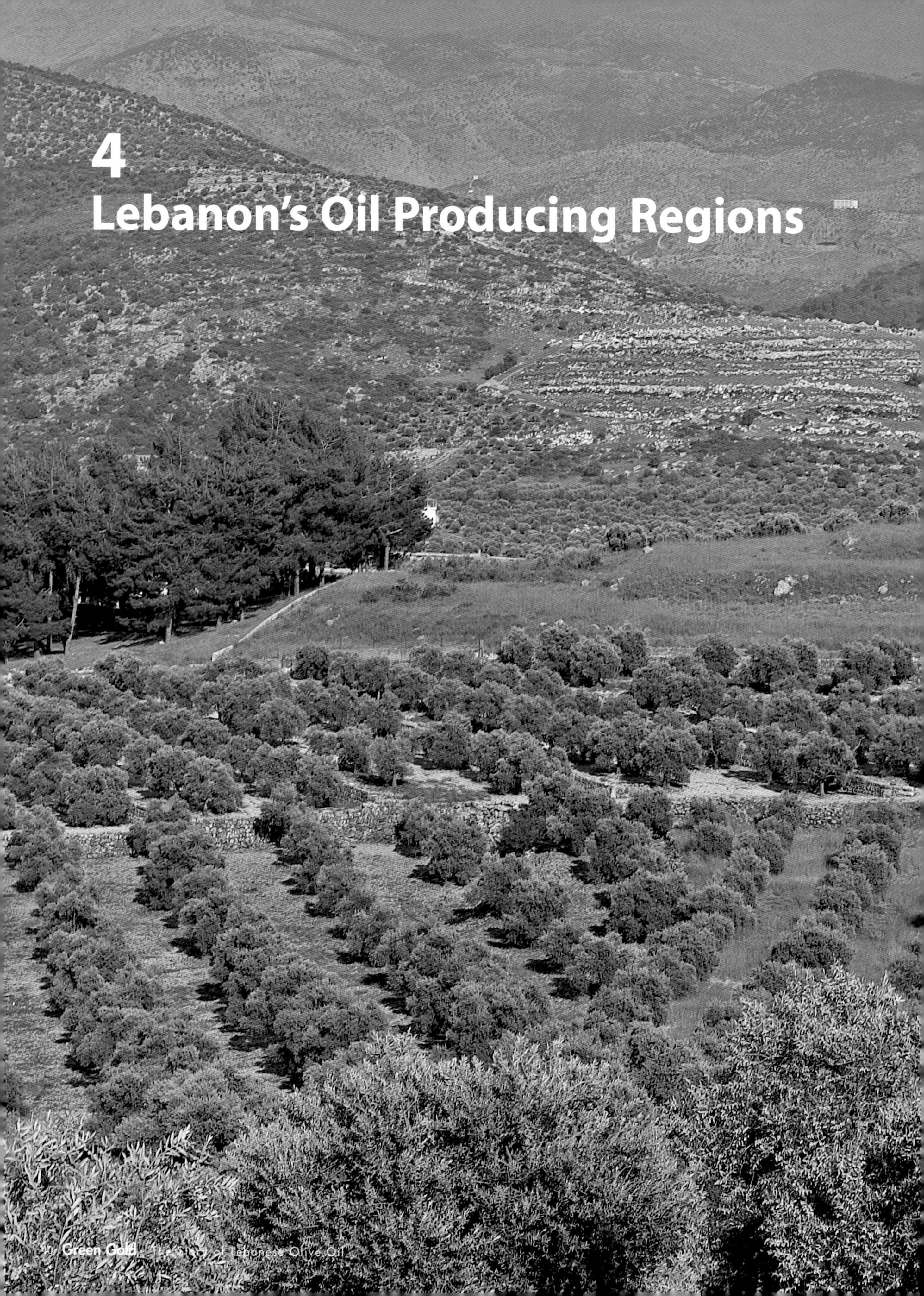

Lebanon's Oil Producing Regions

"His branches shall spread, and his beauty shall be as the olive tree, and his smell as Lebanon"

Hosea 14:6

Left:
Olives have been grown for thousands of years and each region is proud of its olive trees.

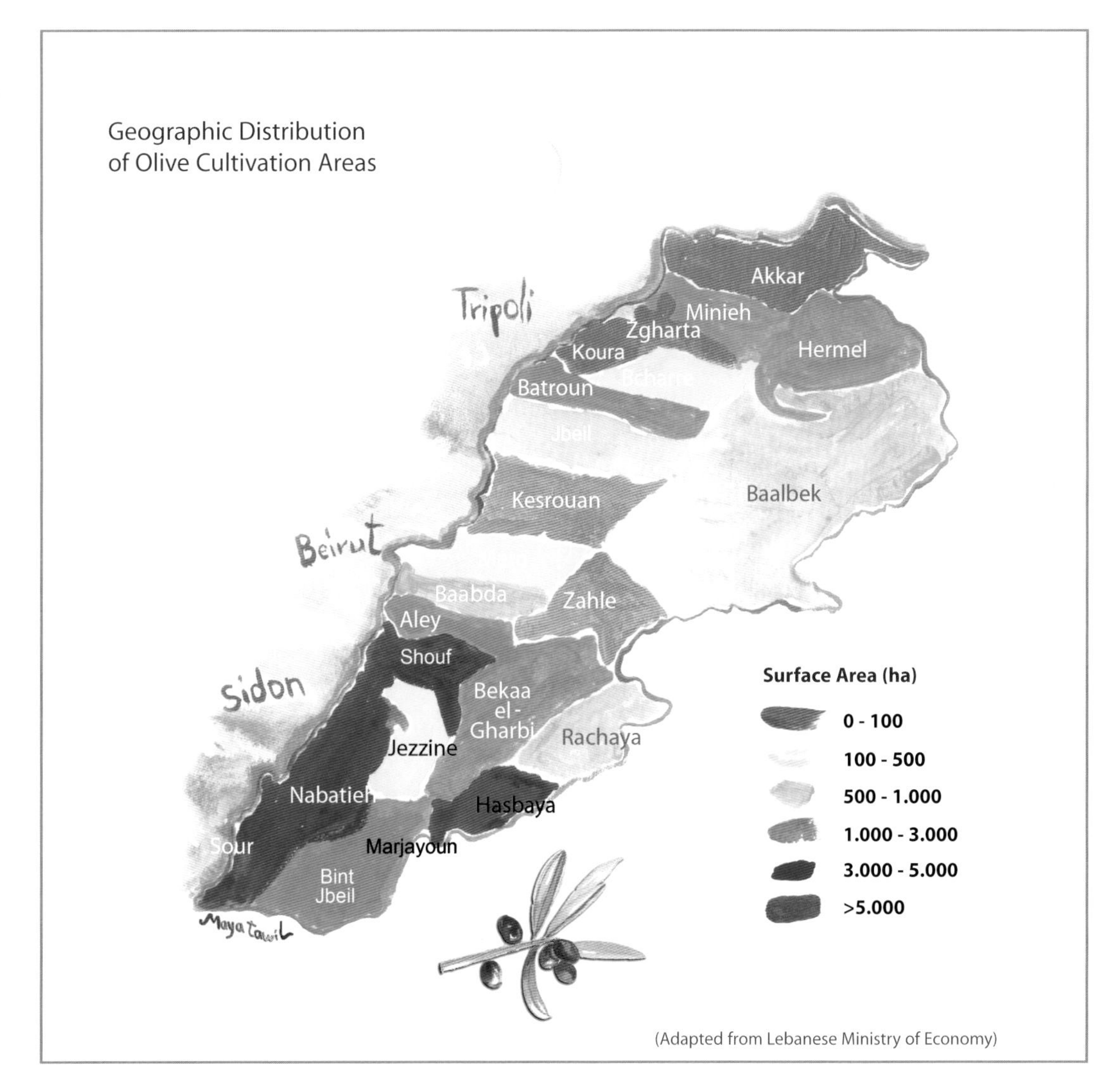

Lebanon's Olive Growing Regions

Olive trees grow in most areas of Lebanon, from the subtropical southern coast to the arid Bekaa Valley and the North, where the mountains are covered in snow in the winter. In fact, North Lebanon, including Akkar, accounts for 60% of all Lebanon's olives and the South, including the Shouf, accounts for most of the rest.

Olive cultivation occupies about 20% of Lebanon's total agricultural land (271,752 hectares). The average field size for more than 50% of Lebanese farmers is less than 0.5 hectares.

The best known oil-producing regions in northern Lebanon are Akkar, Zgharta, Koura and Batroun, while in southern Lebanon, Nabatieh, Hasbaya, Marjayoun, and Rashaya el-Fukhar are most famous. These are all areas where olives have been grown for thousands of years and each area is proud of its olive trees and the oil it produces.

The Lebanese Ministry of Agriculture estimated in 2002 that 57% of all farmers in Lebanon were olive growers. In the last quarter of a century, the area under olive cultivation has nearly doubled, (now 57,000 hectares), sustaining an annual growth rate of around 3%.

Olive growing is the main agricultural activity in many regions in Lebanon, specifically in the North and around Nabatieh in the South, where it accounts for 33% and 47% of all cultivated crops respectively. However, even in areas where the olive is not the primary crop, such as in Mount Lebanon and other parts of South Lebanon, it represents for a substantial percent of total cultivated crops. Olive farming is estimated to be worth around US$200 million, or approximately 20% of the value of total agricultural plant production.

Olive farming is still regarded as a family business and land is handed down from generation to generation. Nowadays, though, it is regarded as a secondary occupation and 70% of olive farmers in South Lebanon, 59% in North Lebanon and 67% in Mount Lebanon, respectively, are only part-time olive growers. Indeed, most of the farmers I spoke to while researching this book have another occupation and many only grow olives for local consumption.

Given its importance, there is a regrettable lack of information regarding the olive industry in Lebanon. Many figures, cited by experts and agencies, are just estimates, or worse, guesstimates. In 2003, only 3% of total production was exported, mostly to the US, Canada and Australia. Since 2003, export sales have quadrupled with the elimination of European Union custom duties on oils, making it easier for Lebanese producers to reach the European market. The increased demand for olive oil worldwide is also a factor.

Most Lebanese farmers still do not feel the need to implement costly international standards, as they prefer to sell their oil to their neighbours, relatives and friends. Furthermore, as local irrigation and pruning methods remain primitive, fluctuations in harvest will inevitably mean that Lebanon is not able to compete with countries using more productive methods, which guarantee more regular supplies of olive oil.

Some farmers though are changing their ways in order to produce a higher quality of oil and some have even obtained organic certification. They realize that they need to comply with international set standards if they are to enter the export market. With the support of initiatives, backed by local NGOs, that have introduced modern oil presses and imposed quality controls, both the quality and quantity of Lebanese oil is on the increase.

Sheikh Sleiman El-Dagher, president of SILO (Syndicate of Interprofessional Lebanese Olive Oil Producers) estimates that 1,150 families in Lebanon make a living from olives. There are 500-odd presses in Lebanon, but only 48 of them extract more than 80% of the country's oil. 85% of the registered mills are considered traditional. The rest are modern automatic or semi-automatic mills.

El-Dagher believes that the way forward for Lebanon is to concentrate on improving the standard of the mills and to focus on producing Extra Virgin olive oil as it fetches the best price when exported. He also envisages a role for the government in funding the planting of olive trees in deprived rural regions.

Lebanon's olive oil industry currently does not enjoy government protection, so cheaper, inferior quality imports are free to flood into the country, often passed off as locally produced oil. This hurts Lebanon's olive farmers and also tarnishes the country's growing reputation as an international producer of quality olive oil.

Many farmers are concerned about what the government can do to protect the local market, so that they can continue to harvest their trees and make an honest living from their groves. There are numerous social and economic implications related to olive oil production in this country. Increased output would provide much-needed work to the rural community, which in turn would help slow down rural migration from remote villages. Olive trees are inherently low maintenance, requiring little day-to-day management compared to other cultivation. This in itself affords farmers the opportunity to engage in other agricultural activities, with the potential to maximize both profits and local employment.

THE NORTH

North Lebanon is home to Lebanon's highest mountain, the 3090-metre Qornet Al-Sawda and two important agricultural areas: Al-Koura and Akkar, both of which are famous for their olives and their fruits. The climate ranges from cold, snowy winters in the mountains to hot, humid summers along the coast.

It boasts Lebanon's second largest city and important trading port, Tripoli. Founded by the Phoenicians in 800 BC, Tripoli is a legend in its own right, and proud heir to historical remains that date back to the Crusades and assorted empires of Islam. It is particularly famous for its old *souks*, traditional market places where tradesmen today still conduct their affairs as they did hundreds of years ago. Tripoli is also famous for its Lebanese desserts and sweets.

The North is also home to Lebanon's most famous writer, Kahlil Gibran, author of *The Prophet*. His hometown of Bsharre is in the al-Arz region, not far from a grove of cedars - some of which are thousands of years old. The grove is a small reminder of the once mighty forests of cedar trees - the symbol of Lebanon - that covered the country and which were often mentioned in the Bible. These trees have been used since Antiquity in construction and shipbuilding.

Beneath Bsharre, the rocky cliffs and precipices of the Qadisha Valley are riddled with grottos and caves and the steep valley walls are studded with ancient monasteries and hermitages. At its western end, the port town of Batroun, which is mentioned in ancient Egyptian records, is a delightful warren of old cobbled roads and traditional Lebanese houses. Best known today for its lemonade, it was here, archaeologists suspect, that the first Phoenician ships were built, the same ships that later carried the olive tree to settlements around the Mediterranean.

THE SOUTH

South Lebanon, a land of citrus, banana and olive groves, has been inhabited for at least 6,000 years. The two main Phoenician cities there, Sidon and Tyre, still exist today. Their monuments, castles and ruins stand as a reminder of their former power and glory.

Tyre is Lebanon's southernmost city and lies just 20 kilometres from the Israeli frontier. It was founded around 3000 BC and eventually became one of the most important Phoenician cities. Originally an island, Tyre was only linked to the mainland during the assault on the city by Alexander the Great, who built a wooden causeway to the island. This silted up and created a permanent land bridge. Its mosaic streets, Roman baths and arena, remains of a Crusader church, extensive Roman-Byzantine necropolis and the largest hippodrome ever built make Tyre a unique and fascinating city.

Sidon, further to the north, was destroyed twice by war between the 7th and 4th centuries BC and once by an earthquake in the 6th century AD. Despite this, the town has many historic places, among them the Eshmoun Temple, an ancient centre of healing and the only Phoenician temple found so far with its carvings and sculptures intact, the extensive remains of a medieval city, complete with bath-houses, marketplaces, churches, mosques and the Sea Castle, a Crusader fortification by the port.

Inland, the Beaufort Castle is the largest and most scenic Crusader castle in Lebanon and has been fought over into recent times.

The largely Christian town of Marjayoun is famous for its beautiful landscape and mild climate and is the regional centre for surrounding villages, which illustrate Lebanon's sectarian mosaic: Sunni and Shiite Muslims, Greek Orthodox and Catholic, Maronite Catholic and Protestant Christians as well as large numbers of Druze, all live in its environs.

Below: *Olive cultivation takes up about 20% of Lebanon's total agricultural land.*

1. NORTH: KFAR AQAA

> *"Zeit emed el-beit"* - *oil is the foundation of the home*
>
> Arab proverb

Right:
Jamil Bou Farah's family has been growing olives and pressing them for oil as far back as he can remember. He grows mainly the Baladi olive variety in his groves in Kfar Aqaa in northern Lebanon.

The north of Lebanon can lay claim to being the main olive growing region in the country and boasts a four-season climate which nurtures the soil. It is one of the most fertile areas in Lebanon. Enveloped by silver-grey hills, which are carpeted in a glorious display of wild flowers each spring, the Koura highlands are legendary throughout Lebanon for their oil. My destination is Kfar Aqaa, a village off the northern highway just past the coastal town of Shekka, near the mountain village of Koura at 400 metres above sea level. It hard not to notice the two olive press stones decorating the roundabout at the turn-off to the village.

It is a warm day in March; snow still lies in drifts on the mountains above al-Arz and fields all around are carpeted in pink arabis, wild crocuses, crown daisies and corn poppies. I am on my way to meet up with Jamil Bou Farah, to share his memories of the olive harvests of bygone days.

"*Zeit emed el-beit*," he tells me: oil is the foundation of the home.

Bou Farah's family has been growing olives and pressing them for oil as far back as he can remember. On his land, he has trees dating back hundreds of years. Jamil is a proud gentleman of the Old School kind, who shows no sign of letting his age slow him down, despite being in his mid-eighties. He has a shop under his house selling seeds and fertilizers to the local farmers and today he is quite busy with customers. He has the vigour of someone half his age.

Once the welcoming coffee formalities are over, he tells me that in the past, his family lived off their oil. There was no soya or corn oil then, so everyone cooked with olive oil and demand was higher. Oil was so valuable that parents of a bride-to-be would even barter oil for her trousseau.

Jamil attended a private school on this income, as did his brothers and sisters, and in 1949 he was even able to graduate from the American University of Beirut. He never intended to work on the land, and harboured ambitions of becoming an accountant. He also did a spot of teaching and, incredibly, still remembers students he taught at a college in Tripoli over 50 years ago.

He recalls that his grandmother advised him never to neglect the olive trees of his ancestors.

"Stick to your roots, she would tell me, they will never let you down."

Today he has 140,000m^2 of olive groves - with around 35 olive trees per 1,000m^2. He knows how fickle the olive can be.

"Last year, we got about 450 *tanke* (containers holding around 18 litres each) and we sold them for $50 each. This year the yield was only a fraction of that."

All over Lebanon in 2005, yields were down, so the price for oil on the market doubled. But with such fluctuations in yield, it is tougher to make your living from olive oil than it used to be, which is why Jamil's children now live and work in Beirut.

He grows mainly the Baladi variety, insisting the skin is stronger and more resilient to harsher climates and disease. His groves are at an altitude of about 350 metres. We sit together and he tells me about a lecture he held at the local school. Never at a loss for words, Jamil seems to enjoy amusing me with proverbs.

"On the back of a tree there is always a new sibling, " he says. "When a king (i.e. the olive tree) dies, there is always another to replace him."

He means that the olive tree does not die: new branches always sprout to replace the dead. He smiles. His bright eyes convey just how much he loves his trees. Jamil follows a basic pruning regime and only ploughs his fields three times a year. He admits this is probably not enough, but says he can't afford to do more.

"Every year we prune the trees and cut away the dead branches, but we are not as thorough as we should be; labour is costly."

As for irrigation, Jamil relies on Mother Nature to send him the rain he needs. He is partly mechanised though. Before 1950, he used oxen to plough the fields; now he uses a tractor.

"Why not use donkeys?" I ask.

"They cannot do the work, they are not strong enough to pull the plough through the soil."

Jamil harvests in November, after the first rains - whatever falls to the ground before this is collected, pressed and the oil is used for soap-making.

Left:
Three generations of olive enthusiasts.

Above:
The Koura highlands are legendary throughout Lebanon for their oil.

The harvest is not done in one go. Depending on the availability of labour, it sometimes continues until February. Migrant workers are usually employed and are paid a daily rate of $10 for the men and $7 for the women.

"Why?" I ask, defending my gender.

"Because the men have to work harder as they need to stretch up further for the picking," he replies.

Olives fall onto a net or canvas spread out beneath the trees to catch falling fruit and await swift collection to avoid oxidation.

The harvest in the area used to be a joyous occasion. Pickers would climb the trees and sing special songs called *ertaba*. There were verses specific to each region.

"We have lost this festiveness," he mourns. "The (civil) war changed all that."

Now the pickers bring a portable radio and instead of harvest songs, the latest hits by Arab heart-throbs echo across the valley.

Jamil owns a small press in the village with his brother, which he purchased from Italy many years ago. He swears by his stone grinding method, insisting that oil made this way will keep for longer.

The process is as follows: the sacks are weighed, and the olives are poured into a bin, washed and then crushed by the stone wheels of the press. In a good year, 100 kilos of olives will yield about 30 kilos of oil, which Jamil sells in tin canisters of 18 or 20 litres to local residents or friends who come from Beirut to buy it.

"Olive oil and onions are the *taj*, the staple, of Lebanese cooking," he says. "Can you imagine a *mou-jaddarah* (lentils and rice) or *loubiyeh* (green beans and tomatoes) without olive oil?"

He leads me down the steep stairs to his cellar without stumbling, even though he walks with a stick, and lets me taste his oil from a canister. I have learnt by now to take a tiny spoonful and not a gulp, when tasting. Jamil's oil is golden in colour and smooth in taste.

"With perhaps a hint of dried almonds?" I ask.

He just nods and asks if I like it.

"Yes, it's very good."

"Come along now, I want to show you my groves."

He drives out through the bushy terrain in a white vintage Mercedes with the enthusiasm of a young-ster taking an off-road spin. We drive through a clearing between the olive groves where the grass is as high as the car. I quickly pull my arm in so as not to get slashed. I am worried about the chassis.

"It's good for the car," he insists, "gives it a good clean."

When I comment on his driving skills, he replies with the old French saying.

"An old pot makes the best soup."

How can I argue with that?

Jamil's olive trees are surrounded by apple, pear and apricot trees. There are not many almond trees here now even though the area used to be full of them years ago. Jamil is not able to work his fields anymore, but he still enjoys the trip to his groves. He proudly tells me he was the first in the area to plant trees that did not bear fruits and points to the tall cypresses on a nearby hill. This part of Lebanon has the most beautiful wild flowers and we take a walk through the anemones, daisies, fox-gloves - I even spot a dark purple flower, the irissofariana, growing under the olives. He invites me to pick some to take home, but I prefer to leave them where they belong.

We head back to visit his olive press, which is now in the hands of his brother. Jamil still enjoys a spe-cial bond with all the farmers who come to the press to deliver their olives. He sits with them during the pressing, praising the green liquid spurting into the containers, insisting that he can judge its

Above:
Father Ksas has turned to olive cultivation to provide locals with a living.

quality by its aroma. The residue from the olive pressing lies stacked in a corner, waiting to be used as winter fuel.

"I used to go to school carrying a clay pot heated with that stuff - we would keep it next to the desk to stay warm in winter," he tells me. "And the teacher's pet would place theirs next to hers to get into her good books!"

There is a trace of regret in his voice when he talks about the past. Jamil knows that his country is falling under the spell of urbanization.

"I feel the happiest here amongst my olive groves," he tells me before reminiscing about a trip to Holland with his wife years ago to choose a diamond for her ring.

"The dealer told me that she only polished the stones with something just as precious as diamonds: olive oil."

2. NORTH LEBANON: BZIZA

"When thou beatest thine olive tree, thou shalt not go over the boughs again: it shall be for the stranger, for the fatherless, and for the widow"

Deuteronomy 24:20

Father Ksas, a pioneering olive tree-growing Maronite Catholic priest, lives in the northern Lebanese village of Bziza, near Amioun in the Koura region. He told me about a local rumour concerning an icon in a nearby church, which miraculously leaks olive oil on occasion.

So on a warm spring morning I head north, taking the highway to Shekka, about 60km north of Beirut on the road to Tripoli. To reach Amioun, I take a right turn off the main road and drive up towards al-Arz, the Cedars. The road to Amioun rises steeply through a gorge, which splits the mountainside. The slopes on each side of the road are marked with olives trees and far away on the mountains, the dark green smudges of the Cedars can be seen bathed in balmy sunlight.

I have been here several times, to visit friends or to bring visitors to see the famed frescoed churches that date back to the Crusader period. It was also up here that I discovered one of the oldest olive trees in Lebanon, apparently over 1,500 years old. This massive olive tree - its size is not immediately obvious as it is lost in a thicket of bushes and wild plants - is on the side of the road below the Mar Youhanna (St John's) church, which is set on a rocky cliff studded with Roman tombs (see page 18). The miraculous icon is housed in the nearby St Georges church.

"I saw oil trickling from an icon of the Virgin Mary during a wedding I attended at the church," Father Ksas later told me.

Left:
The north of Lebanon is the main olive growing region in the country and boasts a four - season climate which nurtures the soil.

As a Maronite, Father Ksas accepts such miracles, but he tells me that the predominantly Greek Orthodox community of the Koura region does not. They do, however, use olive oil in their ceremonies, including marriages and baptisms, and their churches are decorated with icons, some glazed with oil to give them a glossy appearance.

The heavy church door is locked but I come across an old lady in a nearby house who begins to tell me the story of the miracle. I need to press on, I have a meeting with the Father, so I just ask for directions to Bziza.

Armed with instructions to go *yameen, shmeil,* right and left, I head off down the valley. On the way, I pass several olive presses and also a shepherd herding his flock. I cannot resist the perfect picture of baby goats and sheep against the backdrop of the luscious flower-carpeted hills. Somewhat embarrassed, I stop to ask him if I can take his picture. He nods, looking amused, and even strikes a sideways pose for me.

Bziza is home to the ruins of a Roman temple. Its imposing walls are still standing. Somehow, such grandeur looks out of place here. Local inhabitants have planted a little garden around the temple and added a statue of the Virgin Mary. The road to the Father Ksas' house is strewn with red soil. He comes out to meet me and apologizes for the mess, telling me that a lorry dropped its load in the wrong place.

"The red soil was meant for the newly-planted olive trees," he says.

Father Ksas has turned to olive cultivation. He has big plans for the village. He is planting olive trees in the terraces in order to provide locals with a future livelihood. He hopes that one day, there will be enough olives to produce plenty of oil. He says that the villagers are not well off and there are few jobs in this region. The oil would help them enormously.

"I want to educate the youth, to get them started in rural life. The trees will eventually provide an income to help them build homes to stay in the village and get married."

The Father's wife and two children greet me, and I tell him that this is the first time I have meet a Maronite priest who is married. As a husband, Father Ksas understands the difficulties a young couple face when they have no money to buy or rent a home.

Olive growing can be a good business, he tells me, netting as much as a 300 to 400% profit in a good year. Father Ksas has already planted 13,000m^2 and is preparing to plant another 12,000m^2. This year will hopefully see the first harvest and the first olives pressed for oil.

I ask him how he can afford to buy all these trees, and he tells me that a local politician is helping with costs and the church has provided the land.

As a priest, Father Ksas recognizes the importance of olive oil, not just for use in the traditions of the church but also as a source of revenue. He admits to leaving the matters of ploughing and pruning to the farmers, while he focuses on ways to raise money to buy more trees.

He gestures towards the pastoral panorama of olive groves, now wreathed in a sea of mist and uncanny stillness, as the sun sinks into the Mediterranean.

"There is so much fertile land here," he says passionately, "and I want to see it covered with silver-green leaves for as far as I can see."

3. SHOUF: BAAKLINE

"El-zeit zeitoun rayess el-taouleh"

- the olive oil is the king of the table

Lebanese proverb

One cloudy morning, I find myself on the highway south. Passing the coastal town of Damour, I turn off and take the road up to Beiteddine in the Shouf mountains. I love this wide road, which runs through lush, sweeping valleys, past the Damour River and its many little riverside restaurants. I associate this road with trips to the annual cultural and music festival held in the palace of Beiteddine during the hot summer months. Today, I am taking the road to meet an act of a different kind, a farmer-cum-musician and recording artist from the town of Baakline. Everyone calls him *Baba Bia*, Father of the Environment, because of his organic growing practices.

Although the Shouf is just a 20-minute drive out of Beirut, the region has not suffered the creeping concretization of other parts of the country. Picturesque red-roofed houses are abundant on these hillsides of Mount Lebanon and the countryside is unsullied. Druze men wearing *sherwani*, traditional loose fitting black pants gathered at the ankle, and white fezes are a common sight, as are Druze women in black, wearing long, semi-transparent white veils loosely draped over their heads.

A heterodox offshoot of Shi'ite Islam, little is known about the Druze beliefs. Outwardly at least, the Druze are not especially religious and to be Druze, one must be born Druze. Conversion is not possible. Only a few elders, known as the *okal*, are instructed into the doctrines of the faith and it is up to them to ensure its preservation.

Baakline was the capital of the Druze Ma'an princes until the early 17th century, when Emir Fakhreddine II, the most important of them all, moved his capital to Deir el-Qamar. From there, he ruled over a proto-Lebanese state, which stretched at one point far into modern-day Syria. Fakhreddine was able to unite all his subjects and is known for the even-handed way in which he appointed people of all faiths to positions of authority during his rule.

Baakline is also known for its magnificent library, housed in the neoclassical *serail* (municipal palace), which served as the administrative centre for the *mouhafazat* (district) of Mount Lebanon until the Second World War. After a spell as a prison, this building now stocks Arabic, French and English books and publications. It was here that I had arranged to meet *Baba Bia*, or as he is more properly known, Marwan Khoder.

He arrives punctually and as we head out to his groves, Marwan tells me that he has about 4,000m^2 of land at about 650 metres above sea level. The soil is calcareous and grey. These are good conditions for growing olives, as this type of soil retains the moisture from rain and humidity. As we walk across his land, we stop on the way to greet the labourers, who painstakingly till Marwan's narrow terraces by horse.

"This year, the ploughing is taking place late due to unexpectedly heavy rainfall," he explains. "I used to have bulls for ploughing, but when they died they were too costly to replace."

The terraces open into an orchard and we tread through tall grass to a little hut that he proudly tells me he has built out of material salvaged from rubbish tips. He is quite proud of his recycling efforts and for some reason has installed a bell which he just loves to ring at intervals.

Above left:
Beiteddine Palace.

Above right:
The soil in the olive groves surrounding the Shouf village of Baakline is calcareous and grey. These are good conditions for growing olives as this type of soil retains the moisture from rain and humidity.

Right:
Marwan Khoder, farmer-cum-musician and recording artist from Baakline, is known as Baba Bia, Father of the Environment, because of his organic growing practices.

"Would you like some tea?" he enquires and proceeds to light a camping gas fire without waiting for my response. He fills the kettle with water and puts it to boil. Then he walks to the groves and picks different herbs and plants randomly, naming each in turn.

"Mishi, nana ..."

Now what were those others again?

Anyway, he makes a delicious cup of tea with his plethora of herbs and gives me some thyme and *hindbi*, spinach-like leaves, to nibble on.

When it comes to olives, Marwan grows Souri for oil and Baladi for pickling. His tending methods are basic, pruning is done by hand or using a saw. He has never had any specialized training in olive oil production but instead learnt traditions handed down by his forefathers, who have been growing olives for 200 years.

While his ancestors managed to make a living from their land alone, Marwan is forced to look after other peoples' land to supplement his income. He harvests late, usually in December, waiting for his olives to darken. The olives that are left on his trees after the harvest are picked in March and eaten raw from the tree. These olives are known as *jajeer.* The local villagers always help Marwan and his family with the harvest.

Left:
Once picked, the olives are placed straight into a basket - or sometimes an old milk powder can!

Above right:
Many villagers in Lebanon have resisted the winds of change and remain true to their local pressing traditions.

Marwan's trees produce between 25 and 50 kilos of olives and he takes them to one of the 3 presses in Baakline. He stores his oil in clay jars and sells it locally.

He lets me taste his oil. It is smooth, with a hint of almond and golden in colour. Apart from olives, Marwan also grows figs and grapes and is particularly proud of his *zaarour* tree (a type of wild apple). Allegedly the yellow fruit is good for the heart and circulation, and tea can be made out of the leaves.

I ask him about a hiking trail which I took years ago from Baakline down into the valley. I remember that it passed a cave and a natural clear water pool. I would love to do that hike again, I tell him. Apparently, a dip in the water increases longevity. Shame I never tried it out.

"Next time you come, we will not walk, but instead have a picnic," he says, with typical Lebanese generosity.

The clouds close in around us. It's getting cool so he starts to pack up and we head back through the groves. *Yateekon el-aafyeh*, God give you strength, I repeat after him to the labourers. Marwan shows me the cherry trees he has planted. They are an experiment. The soil is calcareous here so not everything prospers, although onions and garlic grow really well. He only uses organic fertilizers, mainly goat dung. I tell him that I am always on the lookout for ancient olive trees. The ones on his land are no more than 100 years old, so he takes me to Said Hemedi's land where there are trees that were planted by Emir Fakhreddine II.

Fakhreddine was born in Baakline. When he became ruler of Lebanon, he moved to the nearby town of Deir el-Qamar before he was forced to flee to Tuscany in 1613, fearing execution at the hands of the Ottomans, who resented Fakhreddine's attempts to be independent. On hearing that his palace in Deir el-Qamar had been burnt down by the Ottoman governor of Tripoli, Fakhreddine returned to Lebanon. The story goes that as revenge, he captured the governor's palace in Akkar and had all the stones dismantled and shipped from Tripoli to Damour, where a human chain of soldiers and subjects carried them up the mountain, to build a new palace.

Fakhreddine is credited with stimulating commerce and agriculture in the Shouf. He brought Italian families to Lebanon to teach new methods of farming and olive cultivation. Thousands of olive trees were planted, and the nearby port of Sidon bustled with ships transporting the oil to Europe.

History stands still in Baakline, in the form of these magnificent olive trees. Many are split in half, as if they have been hit by lightening, but this is due to old age. They stand, almost two separate beings, joined together only by their roots.

I find the largest tree. It is hollow and ancient but it still produces olives every year. It is majestic. I ask Marwan to help me to size it up. We strain our necks and finally agree on a height of seven meters. Then, we walk around the trunk to measure its circumference. When we meet, our joint calculation comes to an incredible 10 metres.

As I leave, I promise to pay a longer visit next time and in turn, I am promised a live *oud* (the Arab lute) performance. Marwan gives me a CD of songs he has recorded, children's songs that celebrate nature, sponsored by the environmental NGO Greenline.

"Are olives mentioned in your songs?" I joke. "Otherwise, I cannot accept your CD."

"Akeed, el-rayess el-taouleh," he replies. Of course, (they are) the king of the table. He waves goodbye as we part.

"I'll be back for the harvest," I say as I drive away.

My Shouf day is not over yet. The pursuit of famous olive oil soap-makers leads me into the narrow side streets of Baakline. I am looking for Sheikh Ali Jubay, who has been the local soapmaker for decades. Finally, I find him and he welcomes me into his workshop. I expect his typical Druze Sheikh attire but I do not expect the sparkling ochre eyes, an amazing yellowish-brown colour with brown speckles, which belie his old age.

According to the Encarta World English Dictionary, the definition of a Sheikh is not just that of the leader of an Arab tribe or village or a senior official in an Islamic religious organization, but also of a handsome and physically appealing man. Well, Sheikh Ali is definitely too old for me, but in his day, he must have been a real catch for the local girls.

What interests me is the way he makes soap. He does it the way his father did, using the surplus olive oil left over after pressing that is unfit for consumption. Cut out into large squares, his soaps are a brilliant white. There is no added fragrance, just the pure smell of olives. Sheikh Ali sells his soap by the kilo, weighing the bars on old brass scales before packing them into plastic bags. He presses a stack of brilliant white soap into my arms. My bathroom shelf is already on the verge of collapse, but never refuse a gift of olive oil soap. Thank you, Sheikh Ali.

Above:
Olive oil is essential in the Lebanese kitchen and all homes have several bottles stashed in a cupboard for drizzling over plates of labneh, a type of strained yoghurt - the staple of any Lebanese table.

4. SOUTH: SINAI

"God is the light of the heavens and earth ... lit from a blessed tree, an olive, neither of the East or West, whose oil is well-nigh luminous, though fire scarce touched it: light upon light"

Koran 24:35

My spirits lift whenever I take a trip out of Beirut to the rural areas. Today, I am heading south, which means a drive along a busy highway with lovely views of the sea and banana plantations. Leaving behind the unsightly concrete buildings hastily thrown up by Lebanese expatriates returned from Africa with their hard-earned *masare*, money, I turn off after Sidon, at Zahrani. The road twists and climbs through hills, past dramatic valleys cloaked in spring flowers, taking me up to the town of Sinai, 250 metres above sea level, where I will meet my first woman olive grower.

Most mountain villages of the south were inaccessible for many years, due to Lebanon's war and the Israeli occupation, which only ended in 2000. I have only recently discovered the region's beauty on spring hikes and it has already become a favourite destination when I am in desperate need of some spiritual healing. Here, Mount Hermon's shadow falls across valleys made lush by the Litani River. The beauty of the south's gorges has to be seen to be believed. No matter how poor and isolated this area may be, its wealth of water from the melting snows flows down from the mountains through springs, streams and rivers, bringing life.

The hills are covered with olive, apple and orange groves interspersed with rows of pine trees. I pass flocks of goats, barely discernable high up on the cliffs, but the roads are deserted. There is barely a soul in sight.

Today, Dr Rami Zurayk, professor of agriculture at the American University of Beirut (AUB), is my guide. He steers me towards a village off the beaten track. Rami is so passionate about olive oil, and about rural life in general, that together with Rania Touma, he set up the AUB's Healthy Basket program (run through the university's Faculty of Agriculture & Food Services). The initiative is intended to help farmers sell their organic produce, vegetables, fruits and olive oil, directly to clients in Beirut, eliminating the middleman and the consequent loss in earnings.

As we pull away from the roar of the highway traffic, leaving the gleaming blue sea behind us, Rami enthuses about one of his farmers.

"You just have to meet Suhaila," he tells me. "She's an incredible woman."

62-year-old Suhaila Bannout - 'the lady with the green eyes' as Rami calls her - has quite a history. She does not come from a tradition of olive cultivation, but fell into it about 30 years ago as a means of making a living when her husband died and she was left to bring up her daughter alone. Tragically, her daughter has passed away since my visit. She was only 42.

Suhaila lives in a tiny home, but she receives me with grace and enthusiasm. I notice she does have light tawny eyes and an unlined complexion that belies her age - thanks of course to all the olive oil she consumes.

She puts out a wobbly table, covering it with a cloth reserved especially for the visitors, and serves us glasses of scalding hot black tea and a plate of her homemade yoghurt, smothered in luscious olive oil. I try it with a piece of *saj* (flat bread). The oil is good, golden in colour and mild in taste, with perhaps a hint of green almonds about it.

Suhaila tends her own small grove, as well as groves belonging to other people. She works hard and lives the meagre life typical of the people of these southern villages, who have lived through civil war and Israeli occupation.

"There is an old Lebanese saying, if in need, do not go to Sinai," Rami says wryly. "This region has always been poor."

Over cups of tea, I ask Suhaila about her pruning methods. Contrary to others I have spoken to, she and other local growers actually do it following the harvest in November, rather than in spring. After they pick off all the olives, the trees are given a once over to rid them of unproductive branches. Due to her unstudied pruning habits, Suhaila's average yield per tree is only 10 kilos.

"So when do you know when it's the right time to harvest?" I ask her. "Do you follow the lunar calendar?"

She has never heard of that before and is amused by my question.

"Well, do you pick when the moon is full then?" I insist. In Syriac, the old language of the Levant, Sinai means 'the way of the moon,' *Darb al Qamar* in Arabic, and I need some unusual myths here.

"No, I just squeeze the olives on the tree, and when they are juicy enough I know it's time," she replies bluntly.

I ask her if she has an olive branch in the house.

"Why?" she asks.

"Well," I tell her, "in Spain there is a saying that keeping an olive branch in the house will make the husband faithful and the wife the master of the home."

Suhaila smiles.

"The true expression of the land can be found in the traditional pressed oils," says Rami.

I suppose what he means is that olive oil in Lebanon reflects a whole way of life. Traditional peasant dishes are still part of daily cuisine, and the Lebanese, who are known for their love of life, have gargantuan appetites.

Rami's words make me think of the concept of the 'terroir,' something that has always intrigued me. It is a French term that refers to the soil, and to the climate and how it influences the taste of what is planted in the soil. Does olive oil taste different when it is planted in rich red soil or dry limestone? Is it affected by the vegetation and flowers growing nearby?

Admittedly, the term is usually applied to wine-growing land, but nowadays some olive experts are using it to describe land where the olives grow as well. Many agriculture engineers still scoff at the

Above right:
In rural Lebanon, the harvest is a family affair - young and old alike join in to reap nature's bounty.

Below right:
In spring the tree begins producing its crop of olives.

Left:
Ancient tree trunks are a familiar sight on Lebanon's mountain terraces, home to some of the oldest olive trees in the world, many of them dating back 1,500 years.

idea that the taste of an oil is affected by where the olive grows or what grows next to it. They will insist that it is the variety, the time of harvest, the quality of the fruit and the method of pressing that determines an oil's distinctive taste. But I still love the explanation given to me by one farmer, Mohammed, who I met walking through a grove in the southern village of Hasbaya.

"What makes our oil taste so special are the fruit trees growing nearby and the bees feeding on the spring flowers," he told me. "Our oil is sweet like the honey the bees produce."

During the harvest in Sinai, picking is done by hand and all available family members are called upon to help. Suhaila's olives are stored in a shed sometimes for up to a week - a fact that will no doubt horrify olive experts - until she has gathered enough to send to the press. Then, one of her relatives, Hasib, sends his donkey to help carry the tattered crates of olives to the local press.

Other farmers spread their olives on their roof to dry before they are taken to be pressed, resulting in a strong oil, pungent in taste. It is not to everyone's taste, but the villagers here like it that way and keep most of the oil for their own use, perhaps selling some to neighbours. The people of Sinai appear unconcerned with the race to raise quality, which every other farmer around the country seems determined to win. They are happy with their oil.

In a good year, they swap olives and oil for fruits grown by other farmers. Otherwise, they sell to former inhabitants, who left for Beirut years ago but who return to stock up on oil for the year. This is typical of the Lebanese, who are proud of their village heritage. Even if they left as children, they still take pride in the olive oil of their ancestral home. It's a good thing too, for this trade is often the only thing that provides people like Suhaila with the meagre incomes they have.

Like many Lebanese, Suhaila has an open house policy. During my visit, neighbours and friends keep dropping by for tea. After accepting an invitation for a lunch of *foul* (broadbeans) with rice, we head out to the olive groves.

Passing her cattle shed, we zigzag down a narrow rocky path, avoiding darting lizards and past an exotic-smelling orange grove in full blossom. We carry on through thigh-high wheat fields and I just have to run my hands over their smoothness while walking, nicking my finger in the process. The scenery is impressive, and rolls on uninterrupted and endlessly for miles in every direction. The terraces are not yet the bleached colour they turn in the summer heat. The surrounding hills are in full bloom with crown daisies, purple pea, bell flowers, corn poppies, wild carnations and seas of spanish broom. Many of them have healing properties, like the *tayyoun,* the inula.

"You crush it and apply the juice over a cut to speed healing," Rami tells me. That's good to know, I think, remembering my finger.

It is noon and although the sun is hot, there is a refreshing breeze. Suhaila points out her olive trees across the decline. They are clustered in thick clumps. The trees she tends are quite young.

"They are around 30 years old," she says, "but in the area, you can still find trees hundreds of years old."

We reach her trees, bathed in the sunshine, and I decide to lie down for a nap under one of them, ignoring the threat of lurking snakes. The solitude and silence of the surroundings is intense and I am only conscious of my contentment. The fresh mountain air and warm fragrance of orange blossoms have suddenly made me sleepy. I wish someone would rub my weary feet with olive oil.

5. SOUTH: SHEBAA

"He who brings his olives from the tree to the stone will become wealthy"

Arab proverb

In Lebanon, you can still find ancient olive oil presses dating back to the Iron Age lying ignored by the road or scattered across hillsides. In other countries, such antiquities would be in museums.

A Professor of Anthropology at the Lebanese University, Dr Moheb Chanesaz is the perfect person to take along on a trip to uncover the past, because he has written a book on the ancient olive presses of Lebanon.

"I will show you where the oldest ones are," he promised after our first meeting in a Beirut café. So, on a sunny morning in April, I joined him on a road trip that took us all the way to Shebaa, in the southeastern corner of Lebanon, on the slopes of the Golan Heights.

This area of the country is notorious for the ongoing political and military dispute over the Shebaa Farms, which are still occupied by Israel. The entire area was off-limits during the Israeli occupation but since they left, it has become possible to visit the village of Shebaa. Moheb assures me with a smile that the journey is safe and that I will not be a target for some trigger-happy soldier.

Right : *During the 15th century water mills were used to extract oil from the olive residue.*

On the way we stop off in Khalde, a town on the outskirts of Beirut, where there are oil basins dating back to the Byzantine period lying abandoned next to a snack shop by the side of the road.

"Don't take any pictures, because there is a military base nearby," Moheb warns.

He promises to take me to a site up in the mountains near the town of Aley, where there are the remains of an innovative olive-crushing basin.

In ancient times, there were several methods used for extracting oil in Lebanon; mortar and pestle, crushing underfoot or by using a revolving roller and a stone basin. In the latter case, the roller was held in place by a horizontal wooden shaft that passed through the hole in the stone. It was stabilized by a pivot, which rotated in a hole in the basin. The advantage of this method was that the roller could be turned by animals.

During the late Bronze Age, the lever and weight press was introduced to Lebanon. In this method, olives were crushed in a stone basin and the oil was collected as it spilled into a shallow ledge at the opening of the basin. The Phoenicians are thought to have introduced such presses to North Africa during their voyages.

The next development was the lever and screw-type press. Instead of stone weights, two parallel beams were attached to the main horizontal beam. A threaded plank was attached with a screw going through it. The screw went into a central socket - a stone basin and was held down with a horizontal lever. This lever could revolve, thus pressing down on woven mats loaded with olives that were placed under the main horizontal beam. This crushed the olives more securely.

Such methods were only able to extract half of the oil in the olive. The rest remained in the residue left in the press. It was not until the 15th century that *makhmera,* water mills, were introduced, and the residue from lever presses could finally be squeezed of its remaining oil.

32
36
11
11
11

Left :
The Shebaa museum opened in summer 2005 with the assistance of the NGO Mercy Corps.

"Back then," Moheb explained, "olive growers would sell their residue to mill owners to extract the remaining oil. In some cases, flour mills stopped grinding grains during the olive season and concentrated just on oil extraction."

Unlike fresh olives, the olive residue could wait for a second pressing without running the risk of spoiling. This type of water mill can still be found in Shebaa.

After passing the town of Nabatieh, we find ourselves in an area of deep scenic valleys blanketed with spring flowers. We drive past Beaufort Castle, perched high on a cliff above the Litani River. This Crusader fortification was built to guard the Kingdom of Jerusalem against attack from the Arabs in Syria - thanks to its expansive panorama, it enabled defenders to keep an eye on the road leading to Damascus. Beaufort was occupied by Palestinian guerrillas in the 1970s and in the 1980s, it was captured by Israeli soldiers. Now it stands as a testament to the liberation of the South and its western wall bears a memorial dedicated to that fight.

We pass olive and orange groves tucked between flowing streams, completely deserted apart from the odd flock of sheep or goats. I ask the driver to stop, as I need to get out of the car for some fresh air. To Moheb's amusement, I get out and breathe in the splendour of this area. As we continue further into the mountains, the road deteriorates and as we climb out of the fields, the valley becomes bleak and stony. The higher up we go, the lower the clouds seem, and I begin to feel a chill that is not only brought on by the cooler air, but by the growing sense of desolation. Lebanon's troubled past with its southern neighbour has exacted a heavy toll upon the area, and this part of the country remains neglected and deserted after years of occupation.

Immediately after the withdrawal of Israeli troops in 2000, efforts were made to attract tourists to the seemingly God-forsaken region. With the assistance of the NGO Mercy Corps, Moheb spent the summer of 2005 here turning an old water mill into a museum. With the help of local folk, the mill was rebuilt and Moheb set up informative panels depicting the extraction process. The mill could be used to extract the oil from the olives; an iron mixer, in the shape of a cross, turned at a high speed, powered by the water of the river.

The museum also features wooden tools, painstakingly restored by Moheb. Visitors can attend a pressing and then relax on one of the woven stools and enjoy freshly baked *saj* (flat bread) prepared by local villagers.

After the tour of the mill, we take a hike up the mountainside beside a running stream. The clouds lift, and with it my spirits, as the sun reflects off the water. Shebaa seems less desolate now, just a bit lonely, waiting again to be rediscovered.

It might seem wishful to think that tourists will ever come here for sightseeing, especially given the massive destruction the area once again suffered during the month-long war between Hezbollah and Israel in 2006, but knowing Lebanon's resilience, Shebaa will eventually bounce back. When it does, visitors to the museum will surely be more welcome than ever.

Below left and right:
Lebanese anthropologist Dr Moheb Chanesaz.

Below middle:
The Shebaa museum features ancient tools used in water mills and olive presses.

6. SOUTH: HASBAYA

> *"If olives are not pressed on the same day they are picked, it is like a bride and groom who do not make love on their wedding night"*
>
> Lebanese proverb

I was told that a group of French olive oil experts had selected the *caza*, or district, of Hasbaya in southern Lebanon for its unique-tasting olive oil. The exceptional array of wild plants here are credited by the locals for their oil's taste and immense flavour. The surrounding mountains protect Hasbaya's olive trees from high humidity, and at 800 metres above sea level, the area enjoys a higher rainfall than the rest of southern Lebanon.

So, off I head again, hitching a lift with a driver called Hassan, to discover Hasbaya's oil. Hasbaya is in southeast Lebanon, in the shadow of Mount Hermon. Here, olive oil has provided a source of living to farmers for generations. We drive by patches of cypress and oak trees and I see for myself why the region is so renowned for its natural beauty.

I have chosen the best time of the year to come; a sunny morning in spring. We pass olive groves surrounded by carpets of tiny pink flax, scarlet anemones, corn poppies, cyclamen, wild cactus and almond trees in full blossom.

Before reaching the village of Hasbaya, I spot a strange sight beside the road - an enormous olive tree decked in red ribbons. I ask Hassan to stop the car to take a closer look. In the hollow of the tree there is a battered brown leather trunk filled with old clothes. I ask the men at a nearby shop what the significance is, hoping against hope for some exciting folk tale or myth about clothes trunks in tree trunks.

"A migrant worker must have left it there," one of them replies disappointingly.

We head on to the village. Hasbaya is famous for its old Lebanese red-roofed stone houses and I thought it would be much smaller. As I have not arranged to meet anyone in particular, I randomly ask for directions to any olive grower. I am directed back down the road to a pharmacy, where the pharmacist tells me I can use the phone to call Rashid Zoueid, head of the village olive oil cooperative. Like many Lebanese olive growers, he has another occupation. Rashid is a teacher at the local school and he agrees to meet me during his break.

He begins by telling me about his father, Mohammed.

"He is 88, and he works from 4am until 6pm tending the olive groves," he tells me. "They are his life."

Rashid's family owns about 25,000m^2 of land and their oldest tree is over 1,000 years old. He tells me about their cultivation methods. They prove to be pretty basic. He does a little pruning and the terraces are ploughed once a year, usually in spring.

They grow mainly the Souri in this area, which is known for producing an oil with a distinctive peppery taste, an herbal aroma and olives with an oil content of about 30%. Rashid still picks olives himself during the harvest, with the rest of his family. They harvest early in October, before the onset of rain. He admits that the yield per tree is usually only about 30 kilos and knows that it could be higher if he irrigated. All the olives are pressed at the village mills.

"There are still some family-owned olive presses in Lebanon, but the majority are owned by cooper-

Above:
Olives and labneh are the feature of every Lebanese table and are devoured during breakfast, lunch and dinner.

atives, as the ones in Hasbaya are," he explains. "The olive farmers of the village buy a press, often with funding from the European Community or USAID, and are together responsible for maintenance and running costs. Then they share the profits from the pressing."

Olive oil is essential to the preparation of *moune*, the food supplies kept aside for the winter months. Balls of *labneh*, strained yoghurt, flavoured with thyme and preserved in oil are a staple of the Lebanese mountain diet. Rashid invites me to his home to taste his *labneh* and oil. On arriving, I realize that his family is obviously not expecting us. His wife, mother and children are all sitting huddled on the kitchen floor watching TV. Even though it is past noon, they are still wearing their nightclothes. The living room next door is probably only used for visitors.

His wife jumps up and takes me to the living room where she offers me Arabic coffee and chocolates. Rashid brings out his olive oil and pours it into a small *arak* glass for me to try. It is so pungent, that I choke when swallowing. I look down and cover my mouth to try and hide my coughing fit.

"Your oil is very good," I tell him.

He looks pleased and smiles. I suppose that this emerald oil is not meant for drinking straight from a glass but rather should be enjoyed poured over a plate of *labneh*, or sliced tomatoes. His wife repeats her offer of chocolates and I am not allowed to leave without taking a few. This behaviour is so typical of the generous nature of Lebanese villagers. A woman I have only just met and who is not expecting to meet me not only plies me with coffee and chocolate, but greets me with an invitation to lunch and dinner. I could not even get out of the door without promising to come back again.

As we head back down towards the coast passing the endless groves basking in the balmy sun, I see some farmers still working the land. Here the soil is calcareous and they till it with a mule and a plough. An old man, wearing the traditional dark baggy pants of a Druze topped by a white shirt and black waistcoat, rests on the roots of an old olive tree, staring into space. He is sitting so still, it is as if he is made of wood himself. I wonder what he is thinking. Could this be Mohammed, Rashid's father? I want to stop and speak to him but we speed past. Perhaps I will come back another day, around 3am, and catch him before he heads off on his daily rounds.

5
Caring for the Body

Caring for the Body

In towns and villages all over Lebanon, olive oil is a dietary staple and it is consumed daily. A bowl of *zaatar,* dried wild thyme, and olive oil is always on the kitchen table for dipping bread into as a snack.

Olive oil is a pure fruit juice and considered a true gift from nature. Hippocrates advocated it as a cure 2,500 years ago and the Greek philosopher Democritus, who lived in 400 BC, believed that a diet of olive oil and honey would ensure longevity.

It is believed that daily consumption of olive oil is indeed beneficial to the health. I have been told that in Lebanon, there is a village where the life span is 97 for men and 98 for women. I am still looking for it, but I can tell you that each village I visited while researching this book has lively folk well into their 90s.

Compared to their European counterparts, the Lebanese have a lower incidence of heart attacks. Olive oil is rich in monounsaturated fats and anti-oxidants, which help prevent the blockage of the arteries and lower blood pressure, thus decreasing the risk of heart problems.

Studies by European doctors indicate that olive oil may even reduce bad cholesterol levels (LDL) when taken daily. Around 15% of an adult's daily calorific intake should come from non-fatty acids, which help increase levels of good cholesterol (HDL) and reduce clogging of the arteries. Other significant components of olive oil are vitamins A1, B1 and E and both Omega 3 and Omega 6 - natural antioxidants which defend against damage by Free Radicals to body cells.

The body can produce most fatty acids, but not Oleic acid or Linolenic acid. Olive oil contains them both. Nutritionists believe that the polyphenols (an anti-oxidant substance found in olive oil) may also help combat cancerous cells. Studies have shown that in Mediterranean countries where consumption of olive oil is high, women have a lower chance of developing breast cancer than in the rest of Europe and the United States. Olive oil also protects against the formation of blood clots, stimulates the liver and bile tract and prevents stomach ulcers. Perhaps the Lebanese are right to credit their daily dose of oil for their comparatively good health.

In the past, olive oil was kept at the village pharmacy. These days, every single Lebanese household has at least one bottle of oil in the kitchen cupboard. A villager I once met told me the Arab saying that, "if you enter a friend's house carrying olive oil, make sure you give some away before leaving." In other words, share what is precious with others.

Left:
Use olive oil to rub over hands to soften dry skin.

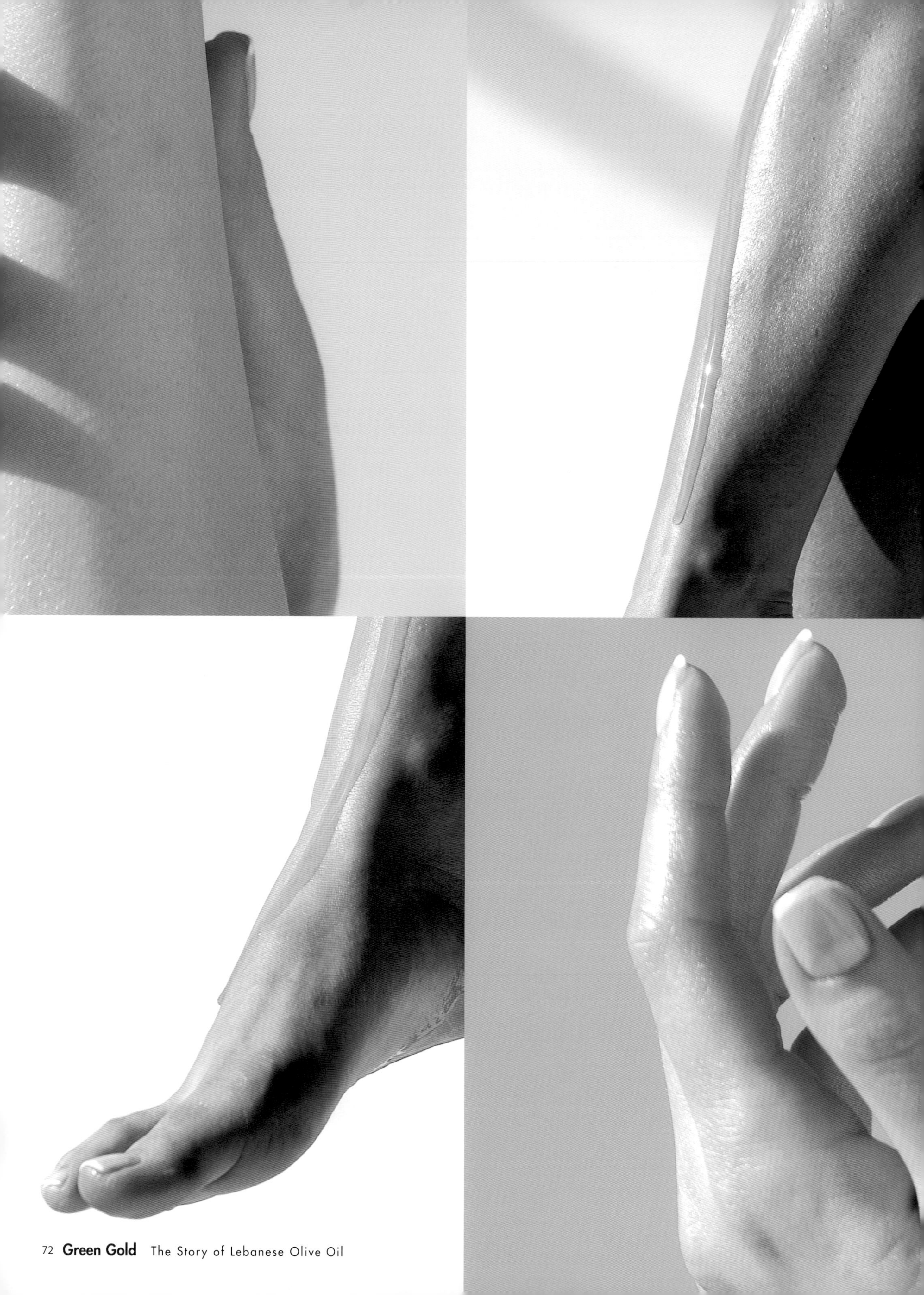

Healing the Body:

Try these tips handed down from generation to generation:

- A tablespoon a day on an empty stomach to combat constipation.
- Figs crushed with olive oil to cleanse the stomach and colon.
- Half a cup of oil mixed with a couple of spoons of lemon juice and a cup of organic honey (warmed over a low heat for a few minutes), to help against a sore throat or a cough. Take a tablespoon every hour.
- A couple of spoons of oil before drinking alcohol lines the stomach and prevents a hangover.
- Massage sore muscles with olive oil after exertion.
- Make a compress of warm oil and herbal tea to apply to blisters.
- To clear spotty skin, dab the area with olive oil mixed with a few dabs of rose water.
- Rub over gums to prevent diseases and teeth to help maintain the whiteness of teeth.
- Rub a spoon of warm oil over the stomach to relieve aches and pains.
- Soak cotton wool with warm oil and place into the ear channel to relieve ear pain.
- Applied topically, olive oil can soothe everything from burns to jellyfish stings. It will also ward off mosquitoes.
- To heal sunburn, or a burn of any kind, mix equal parts of water and oil, beat until thick and then apply it on the skin.

Left and below: *Olive oil has played a role in the Mediterranean beauty routine for centuries.*

The Virgin Body

Right:
For a nourishing hair mask, warm a cup of oil, massage it into the scalp and comb through hair.

The people of the Mediterranean have been using olive oil as part of their beauty regime for centuries. The ancient Egyptians were among the first to benefit from olive oil, which the Canaanites shipped from Lebanon in amphorae that still turn up in ancient shipwrecks. They did not consume the oil, but used it for funeral rituals and cosmetic purposes. In both instances, it was often scented with flowers or herbs.

Today, virtually every Lebanese home has a bar of olive oil soap in the house - some more refined than others - for bathing and for household cleaning.

Tips for face and body

Age-old remedies are often the best. Try the tips below, passed down from generation to generation, to get the most out of Lebanon's Extra Virgin olive oil, which is loaded with antioxidants and is excellent for skin and hair.

Skin

- For radiant skin, take a tablespoon of olive oil first thing in the morning or just add the same amount to your daily diet.
- For exfoliating dry skin, mix a cup of oil with a teaspoon of coarse sea salt. Rub it over the body and then shower off. This is also good for treating ingrown hairs.
- To make a body cream, mix 3/4 litres of olive oil with 1 1/2 ounces of beeswax and heat slowly until it melts. Cool and keep in a covered jar and use for softening dry skin.
- Add a drop of musk essence to few tablespoons of oil. Warm a little between palms of hands and use for a sensuous massage.
- Mix a tablespoon of oil with a few drops of lemon juice, apply to the face and leave overnight to combat wrinkles.
- For a face mask, mash 1/4 of a ripe avocado with a cup of oil until it is well blended, smooth over face, leave for 10 minutes and then rinse off.
- Put a few drops of oil on cotton wool to remove eye make-up.
- Add 4 tablespoons of oil to a warm bath for soft skin all over.
- Mix a couple of tablespoons of oil with 4 drops of peppermint oil for an invigorating foot softener. Massage over feet and then put on some socks to absorb the oil.
- Strengthen nails and soften cuticles by soaking fingernails in a small bowl of warmed olive oil for a few minutes. Use the leftover oil to rub over hands to soften dry skin.
- Mix oil with an equal amount of cocoa butter and apply to stomach, bottom and breasts to prevent stretch marks during pregnancy. Once the baby arrives, keep the oil close by to rub over the scalp to remove cradle cap.

Hair

- As a quick remedy for softer hair, massage with a mixture of an egg yolk, a tablespoon of olive oil, 1/2 a glass of beer and the juice of a lemon. Rinse with warm water and shampoo as usual.

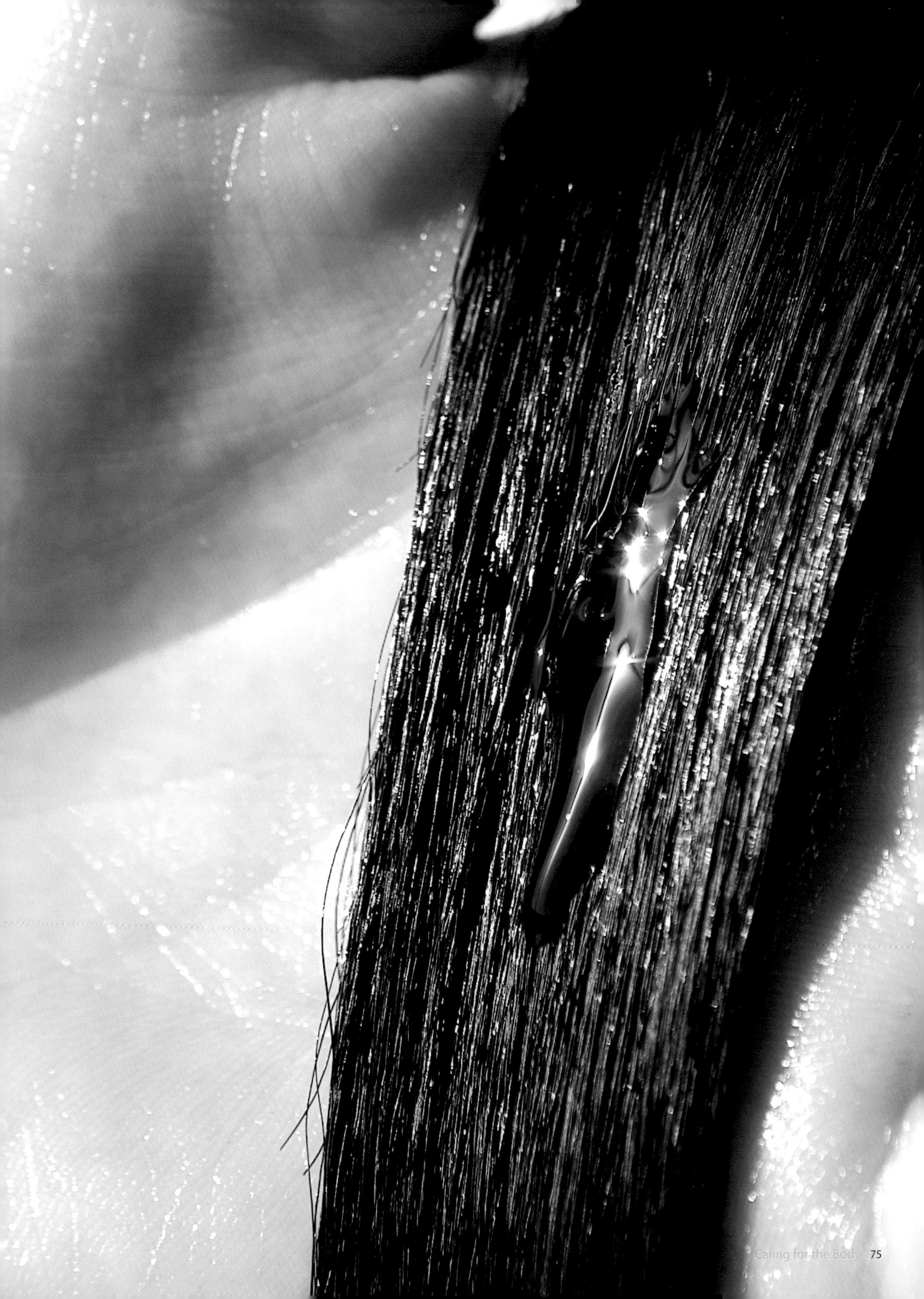

- For a nourishing hair mask, warm a cup of oil, massage it into the scalp and comb through hair. Wrap hair with a towel and leave mask for 10 minutes. Remove towel, rinse and shampoo.
- For frizz control, put a drop of oil in your hand, rub hands together and smooth over hair.
- For a smooth shave, try using oil instead of shaving cream or foam.

Pet Care

- To prevent shedding and to maintain a gleaming coat add 1/2 a teaspoon a day to the dog feed. (1 to 2 spoons for larger dogs.)
- Smother a tick by applying olive oil with a cotton tip and then lift it off slowly from the skin.
- Soften cracked paws with an olive oil massage.

Tips for the home

- Prolong the life of cut flowers by soaking the stems in olive oil before putting them in water.
- Bring back shine to furniture by mixing 4 parts of vinegar with 2 parts of olive oil and 3 parts of turpentine. Rub this mixture on with a soft cloth.
- Put a small amount of oil on a dishwashing towel and use to polish the blade of kitchen knives to prevent rusting. Also polish the handles for extra shine.
- After sweeping the dust from the floor, add a few drops of olive oil and lemon juice to a bucket of water for a final mop. It leaves hardwood and tiled floors gleaming.
- To remove paint from skin, put some oil on a cotton cloth and rub over area.
- Wipe the leaves of indoor plants with a cloth dabbed with oil to make them shine.
- To help pearls and diamonds regain their gleam, dab a little oil on a soft cloth and buff.
- Put a drop of oil onto squeaky door hinges to get rid of the noise.
- Use a drizzle to grease snowboard bindings and the wheels on skateboards.
- Put a tiny drop of oil onto a stuck zipper to loosen, but take care not to get any oil on the fabric itself.

Above:
The southern Lebanese port town of Sidon boasts one of the few soap museums in the world, the Audi Soap Museum. It is housed in a beautiful restored former soap factory.

Saboun Baladi: Traditional Olive Oil Soaps

The last few years have seen a revival in demand for olive oil soaps. Olive-based soaps are one of the most natural cleansing agents for the skin and are renowned for the purity of their ingredients. Although the Phoenicians introduced the olive tree to France, and even though olive oil soap has been produced in the Mediterranean region for millennia, it was not introduced to Europe until the Middle Ages, when Silk Route traders returning to Europe from the Lebanese port cities of Sidon and Tripoli bought the first shipments of soap with them.

Olive oil soap is still made in the traditional way in Lebanon, generally from olives deemed unfit for consumption. However, the best quality soaps are made from Extra Virgin oil.

The process is relatively simple. Oil is mixed with lye, water and glycerine (for added moisture) in large vats and heated. The resulting soap is generally white in colour, although crushed olive pits can be added to the mix to turn it green and beauty soaps are often perfumed with natural fragrances such as rose or lavender. Once cooked, the mixture is poured onto huge moulds on the floor and left to cool.

Before the soap is fully dry, it is scored into cubes, to make it easier to cut later. The rope used in the scoring process is sometimes soaked in coloured dye, which leaves vibrant patterns on the soap. Finally, it is cut into cubes, stacked and left to harden.

Olive oil soap is used not only for personal hygiene but also used for hand-washing clothes. The water used to wash the floor after soap making is highly sought-after by nearby residents who use it to clean their own floors.

Left:
Olive-based soaps are the most natural cleansing agents for the skin and the last years have seen a revival in demand for traditional Lebanese olive oil soaps.

Soaps are usually stamped with the name of the soap maker, or a company logo. Each soap maker - and it is usually a family business in Lebanon - works with jealously-guarded recipes handed down from generation to generation.

As demand grows, olive soap is increasingly being made on a commercial scale. Lebanese companies are producing ever more sophisticated varieties of olive soap, incorporating new blends of oil and new, exotic fragrances to create soaps that are sold at high-end stores and specialist boutiques all over the world.

Audi Soap Museum, Sidon

The ancient city of Sidon, which is now the regional capital of southern Lebanon, began life over 6,000 years ago as a Phoenician city-state. By the Middle Ages, it had become an important regional centre for olive soap manufacturing and until quite recently, Sidon had many functioning soap factories. Today, most have stopped working but the city can now boast one of the few soap museums in the world.

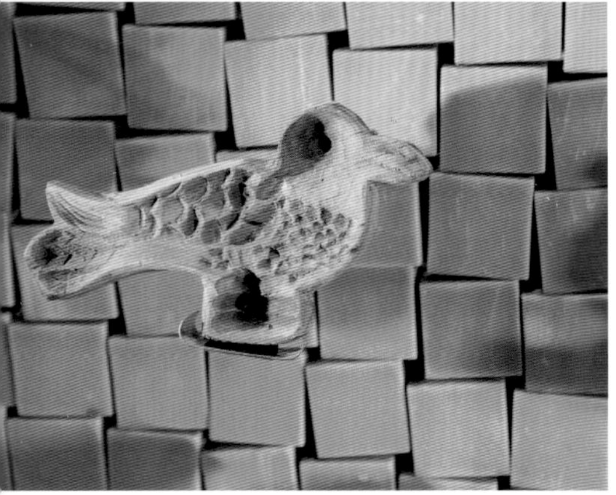

Right:
Soaps are stamped with the name of the soap maker. Many work with jealously-guarded recipes handed down from generation to generation.

Housed in a beautifully-restored former soap factory, the museum was created by Sidon's Audi family, a significant Lebanese banking dynasty. The building itself dates back to at least the 17th century.

Many of the elements of the former factory have been retained and clever displays document the use of olive oil soap throughout the ages. Visitors can also watch a documentary film explaining the different steps of production. The tools once used in soap making are also on display, alongside other items like bathing accessories used in the old *hammams* (Turkish baths).

The museum has a charming café and a small boutique: and after learning about the history of olive soap, visitors can buy cool, white blocks scented with laurel, lavender and amber.

Khan Al-Saboun, Tripoli

Tripoli's Khan al-Saboun (Soap Depot) was built at the beginning of the 17th century by Yusuf al-Saifi, the Ottoman governor of Tripoli. The graceful arcaded stone structure with a large interior courtyard originally served as a barracks for Ottoman troops. Later, the building was turned into a soap factory and storehouse. It is in this archaic setting that the Hassoun family makes traditional Tripolitan olive soap today, as well as a range of olive-based moisturizing creams.

Their signature soap is a multicolour marvel, created by running a rope dipped in various dyes through the soap as it sets. It is moulded into a rough ball that is then polished using a special blade until it is round and smooth. This exotic soap, often scented with aromatic oils such as musk, amber, vanilla or ginger, is frequently given as a gift to Lebanese brides.

Other simpler soaps made here contain honey and shredded mint and are wonderfully refreshing on the skin. Hassoun is best known for a soap assembled in the form of a string of worry beads, which is exported all over the world.

Senteurs D'Orient

The soaps here are significantly more upmarket than the traditional Lebanese versions. The smooth, sophisticated bars are carved into intricate designs and come in attractive packaging. They sell extremely well on the export market.

The olive oil used in these soaps is enhanced with Palm oil and Shea butter, which add extra moisturizing power. For some soaps, roses and laurel are blended into the bar, giving off intense fragrances. Others have added wheat germ or oats, which give them a slightly rough feel, but promote deep cleansing and exfoliation.

Zahia Abboud Soaps

Zahia Abboud makes traditional Lebanese soaps with her own olive oil, to which she adds lavender, musk and laurel. Wrapped in sheer white paper, exuding exotic aromas, the bars are stacked high in her tiny shop, next to her traditional stone olive press. A local artist makes minute sculptures out of her soaps, as well as strings of perfectly shaped worry bead chains. Old family machinery and tools are also on display.

Lebanon's olive oil … looking forward

The olive groves are Lebanon's heartland.

Lebanon's hills and valleys are carpeted with luscious olive groves, and it would be impossible to imagine the country without them. These trees do far more than enhance the beauty of the countryside: the olive has provided the Lebanese with a living for centuries and olive production is an integral part of village society and culture.

As my ramblings through the olive-growing regions draw to a close, I find myself faced with an overwhelmingly positive picture - recent years have seen many new groves being planted in the countryside, especially in the Akkar region of northern Lebanon and throughout the South, following the withdrawal of Israeli troops in 2000.

Sadly, the farmers of southern Lebanon once again suffered a severe setback during the month-long conflict of summer of 2006 between Hezbollah and Israel. Olive groves were turned into death traps by the tens of thousands of cluster bombs scattered by the Israeli army before their withdrawal, making a normal harvest that year almost impossible. As a result, olives were left to rot on the trees by farmers understandably unwilling to risk life and limb for a harvest. The road to recovery will be long and hard, but Lebanon's farmers are resilient and so are their olive groves.

The magnificent oils produced the length and breadth of Lebanon remain largely unknown outside the country. They are an asset that remains to be exploited, overshadowed perhaps by the political turmoil that has dogged the country in its recent history. After thousands of years of cultivation, Lebanon's olive oil, essential to making culinary delights and luscious soaps, remains a treasure to all the country's inhabitants and they continue to take it with them on their journeys to far-flung shores, just as their Canaanite and Phoenician predecessors did.

The olive is tenacious. It thrives through both adversity and good fortune. It requires little, yet affords so much to those who encounter it. It brings health and prosperity. Its self-reliant and self-preserving nature has ensured it success from ancient times to the present day. Little wonder then, that I end my journey pondering the similarities between the olive and the people of Lebanon who, like the olive, treat adversity as a challenge and continue to thrive, however harsh their environment.

I hope that this story will have given the reader the desire to get to know and to enjoy Lebanese olive oil as I have. I also hope my personal odyssey, along the many and convoluted paths of Lebanon's olive groves, and my encounters with the people who dedicate their lives to these generous trees, will afford the reader a different view of Lebanon, one quite different to the one so widely reported in the media. I hope, finally, that it may even shine a light on some of the many possibilities that lie dormant in this magnificent country.

Every road in Lebanon leads to an olive tree.

introduction to lebanese cuisine

Olive oil is at the very heart of Lebanese cuisine. The velvety texture of the oil provides an ideal base for a number of recipes, but it is also used for frying, dressing salads and garnishing – or simply as a dip for freshly-baked bread. Olive oil provides a distinctive flavour of its own, but it also allows the aroma of herbs and spices to develop. Put simply, this natural "green gold" is one of Lebanon's greatest treasures.

Lebanese cooking is tied into the land and its people. Vegetables, grains, and herbs growing in abundance along the coastal plains or in Lebanon's fertile Bekaa Valley are skillfully blended by local housewives, with lamb raised on the sweet mountain grass and fish fresh from the Mediterranean, to provide a healthy, nutritious and economical cuisine. For the Lebanese, cooking is both an art and a pleasure; it is about sharing the abundance of nature with those you love; it is about tradition, with treasured recipes passed down from mother to daughter.

A typical Lebanese meal starts with a selection of mezze (hors d'oeuvre or starters), small plates of dips, salads, pastries and other savory treats, spread out across the dining table. Eating mezze is a convivial experience: conversation flows as locals use bread to scoop up mouthfuls of dips and salads, passing dishes along with gossip and favorite stories. After the mezze, a main course is served. This may be a hearty stew accompanied by rice, barbecued skewers of meat, or a steaming platter of baked or fried fish. A refreshing selection of fresh fruit or delectable Lebanese sweets, served with cups of bitter coffee, provides the exquisite final note to the meal.

We invite you to discover the versatility of olive oil for yourself. Try out some of the following recipes in your own kitchen, and indulge in the flavours and tastes of Lebanon.

the lebanese larder

The following ingredients can be found at Middle Eastern supermarkets and groceries.

Arabic bread: A pocketed flat, round bread, similar to the Greek pita bread.

Burghol (cracked wheat): A wheat grain that has been boiled, dried in the sun and finely ground.

Chickpeas: Round, pale yellow beans, a staple of Lebanese cuisine. Dried beans should be soaked in water for at least 12 hours before cooking.

Dibs remmane: A thick sweet-and-sour sauce made from boiled pomegranate juice, used to flavour savory dishes.

Fava beans: Brown beans, another staple of Lebanese cooking. Dried beans should be soaked in water for at least 12 hours before cooking.

Grape leaves: Where fresh leaves aren't available, grape leaves preserved in brine can be bought from Middle Eastern or Greek groceries.

Kama: Desert musrooms, with a pleasing nutty flavour. Kama can be expensive, and are considered a treat for special occasions.

Okra: Also known as lady's-finger or gumbo, okra is a bright green pod used in many stews and other Lebanese dishes.

Purslane: A wild plant with soft green leaves, Purslane is available at vegetable markets in Lebanon during spring and summer.

Pomegranate seeds: The bright red seeds of the pomegranate are often used as a garnish.

Sumac: A lemony-tasting, coarsely ground powder made from the dried berries of the Sumac plant, which grows wild in the Lebanese mountains.

Tahini: A white, slightly bitter paste with a silky-smooth consistency, made from raw sesame seeds.

Tarator: A creamy white sauce or dip, traditionally served with fish, made from tahini paste and lemon juice.

Equivalent terms

Beets	Beetroot
Eggplant	Aubergine
Okra	Gumbo, lady's-finger
Scallions	Green onions, spring onions
Squash	Courgette, zucchini

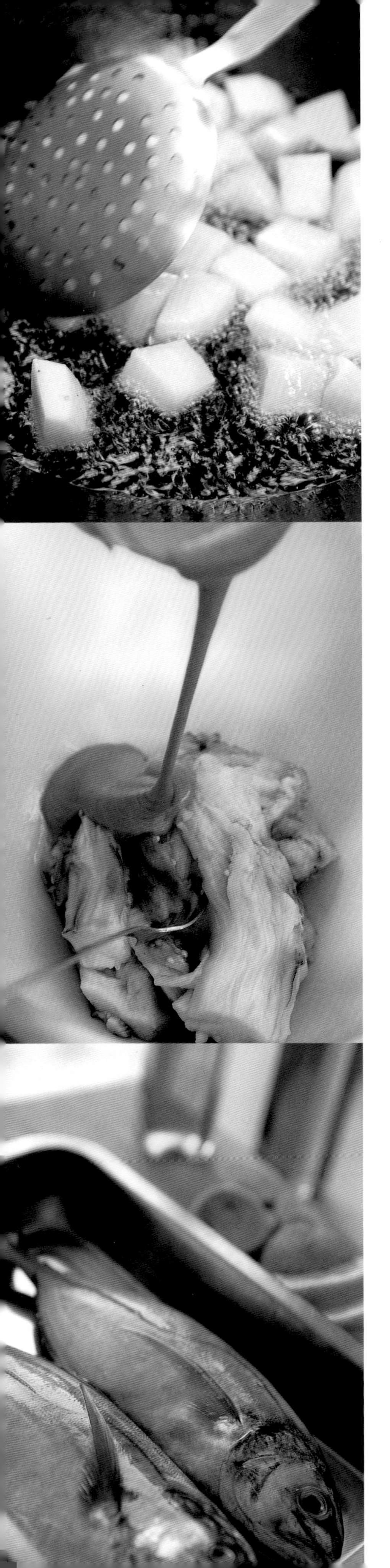

recipes

Mezze

Pickled olives: green and black

Kabees zeitoun

A staple of the Lebanese diet. Most Lebanese serve olives with every meal.

500 grams green olives
500 grams black olives
Lemon wedges

Brine:
2 tablespoons rock salt per cup of water

Green olives are picked before they ripen. To prepare green olives, wash and then soak in water for 2 days - this helps remove any bitterness. Change the water at least twice a day.

Place the olives in sterilized jars with lemon wedges and cover with brine. Add a little olive oil on top, and seal. Leave them at least for 1 month before eating.

To eat freshly-picked green olives, crack them with a mallet and coat them with a little salt - the cracking of the olive also helps remove bitterness. For a mouth-warming delight, try fresh green olives with buttered Arabic bread.

Black olives are picked at harvest time. Rub with coarse salt, cover with water and set aside for about 4 days. Place the olives with lemon wedges in sterilized jars and cover with brine. Add a little olive oil on top, and seal. Leave for at least 1 month before eating.

Chickpea dip

Hommos

An indispensable part of a Lebanese mezze. Everyone loves the smooth, silky texture of hommos.

1 cup chickpeas, cooked
2 cloves garlic, crushed (optional)
1/4 cup tahineh
1/4 cup lemon juice
Olive oil
Salt
Chopped parsley to garnish

Soak 1 cup of dried chickpeas in at least 3 cups of water overnight. The next day, drain and rinse the chickpeas and cook them in a saucepan with at least 5 cups of water for 1-1 1/2 hours, or until tender (to reduce cooking time, use a pressure cooker). Drain the water. While the chickpeas are still warm, blend in a food processor until creamy.

Add the tahineh, crushed garlic, lemon juice and salt to the food processor and pulse to make a smooth paste. If the paste gets too thick, add small amounts of water until you get a creamy consistency.

Hommos should be served at room temperature in small plates garnished with olive oil. You may wish to decorate with some chopped parsley. The dip is eaten with Arabic bread.

Serves 2

Eggplant dip

Moutabel

Also known in Lebanon as *baba ghannouj,* this is an easy and nutritious dip that looks stunning garnished with ruby red pomegranate seeds.

1 medium eggplant
1 clove garlic, crushed
3/4 cup lemon juice
1/8 cup tahineh
3 teaspoons olive oil
Salt
Pomegranate seeds to garnish

Pierce the skin of the eggplant with a fork. Cook it under a hot grill for approximately 15 minutes, turning it half way through to expose both sides to the heat. (When I'm in a hurry, I grill the eggplant on the top burner of a gas stove.) The skin should be blistered and the flesh of the eggplant shriveled and soft to the touch.

When the eggplant is cool, remove the skin and mash the flesh into a pulp. Pulse the pulp with the crushed garlic, lemon juice and tahineh in a blender or food processor to make a fine paste. Spoon the mixture into a small bowl or bowls and pour olive oil on top.

Garnish with pomegranate seeds, and serve at room temperature with Arabic bread.

Serves 2

Fava beans in lemon juice, garlic & olive oil

Foul modammas

A simple, sustaining dish, frequently eaten at breakfast.

1 cup fava beans, cooked
2 cloves garlic, crushed
2 tablespoons lemon juice
2 tablespoons olive oil
Salt

Wash the dried fava beans and leave to soak in at least 3 cups of water for 12 hours.

After soaking, rinse the beans and place in a saucepan with at least 5 cups of water and cook for about an hour, or until tender. Drain and place in a mixing bowl. Lightly mash the beans, adding the crushed garlic and lemon juice. Spoon the mixture into a small bowl.

Pour olive oil over the beans and serve warm or at room temperature with Arabic bread.

Serves 2

Fava beans with chickpeas

Foul w hommos

A variation on the foul recipe, which marries the tastes and textures of cooked chickpeas and fava beans, both nutritious and high in protein.

1 recipe of foul (see opposite page)
1/4 cup chickpeas, cooked (see page 93)

Prepare the recipe for foul. Spoon the foul into a small bowl and add whole (uncrushed) cooked chickpeas in the center. Pour olive oil on top.

Serves 2

String beans in olive oil

Loubiyeh bi zeit

Best eaten the day after (if there is any left!) to give the flavour time to fully develop. Stringing green beans has a calming effect - almost a form of culinary meditation.

500 grams string beans
2 large tomatoes, diced
2 cloves garlic, crushed
1 large onion, diced
1 1/2 tablespoons tomato paste
1/4 cup olive oil
Salt and pepper

Top, tail and string the beans and cut into 2-inch pieces. Dice the onions and crush the garlic.

In a 2-quart saucepan, heat the olive oil and sauté the string beans and diced onion until translucent. (Always heat olive oil before you add any vegetable, in order to avoid heartburn.) Next, add the diced tomato and crushed garlic.

Dilute the tomato paste in about 3 cups of lukewarm water. Add enough of the liquid to cover the stew, and season with salt and pepper to taste. Cover the saucepan and simmer on low flame for about 40 minutes, or until the beans are tender and the sauce has thickened.

Serve at room temperature.

Serves 2 to 4

Potatoes with garlic and coriander

Batata bi kouzbara

The flavour of the coriander lifts the simple potato to another culinary plateau.

500 grams potatoes
3 cloves garlic, crushed
1 cup coriander, chopped
1 cup olive oil
Fresh lemon juice
Salt and pepper

Peel the potatoes and cut into 1-inch cubes, then fry in olive oil until golden brown. Remove the cubes from the frying pan, and drain on paper towel. Add the chopped coriander and crushed garlic to the same oil and sauté until limp.

In a large bowl, combine the potatoes with the coriander mixture and squeeze about 1/2 cup of fresh lemon juice over the dish immediately before serving.

Serves 2 to 4

Cream cheese

Labneh

A delicious spread made from strained yoghurt.

1 recipe of homemade laban (yoghurt), or 4 cups of store bought yoghurt

Salt

Mix the laban with 1 tablespoon of salt and pour into a strainer lined with paper towels. Set over a large bowl to allow the water to drain out. Place in the refrigerator for at least a day.

Then remove the soft spread (labneh) with a spoon into a covered container. Pour some olive oil on top, to prevent a dry crust from forming.

Spread the labneh on a small platter and drizzle some olive oil on top. Labneh is traditionally served with fresh cucumbers, sliced tomatoes, olives and Arabic bread at breakfast or as a mezze dish.

Serves 2

Recipe for homemade laban:

1 quart of milk

(2 pints or 1.2 litres)

1/4 cup of starter (laban made from a previous batch of yoghurt. If the starter is 'new,' the laban will be sweet; if it is 'old' it will be a little tart.)

Boil the milk, then lower heat and allow to cool. When the milk is lukewarm, add the starter, then mix. A good way to test if the milk has cooled down sufficiently is to dip your pinky into the milk: if you can hold it there until the count of 10, it is ready.

Cover the saucepan of milk with a lid, and wrap a blanket or a wool sweater around it. Store in a warm place. Let it stand undisturbed for at least 6 hours so it can set. Then place the new laban in the refrigerator until you wish to make labneh.

Desert mushrooms with onions

Kama

Kama is a wild mushroom - similar in shape to a truffle but softer in consistency - which sprouts in the desert sand after a thunder and lightening storm. They are available in Lebanon from November to February.

500 grams kama
1 large onion, diced
Olive oil

Thoroughly wash the kama, peel and thinly slice.

Fry the diced onion in hot olive oil for a few minutes until translucent. Add the kama and sauté very briefly - just long enough to coat the mushrooms.

Serve at room temperature with lemon wedges. This recipe is especially good served over a fried steak.

Serves 2 to 4

Salad

Parsley, tomato & cracked wheat salad

Tabbouleh

To me, Lebanon and tabbouleh are synonymous! The secret to good tabbouleh is patiently hand-chopping the parsley and mint into thin strips.

2 cups flat parsley leaves, finely chopped (equivalent to 2 bunches)
1/4 cup burghol (cracked wheat)
1 medium onion, diced
1 cup tomatoes, diced
1/4 cup fresh mint leaves, finely chopped
1/2 cup lemon juice
1/2 cup olive oil
Salt and pepper
Grape, lettuce or cabbage leaves

Wash the parsley, divide into small bunches and leave to drain (I usually do this the night before I plan to make tabbouleh). If you want to make tabbouleh right away, wash the parsley and leave to dry for at least 1 hour. Pat with paper towels to absorb any additional moisture (if not completely dry, the leaves will be slimy). Once the parsley is dry, finely chop and set aside.

Meanwhile, dice the onions very fine, rub them with salt and pepper and set aside.

Rinse the burghol until the water runs clear and drain, squeezing any excess water out with your hands. Place the burghol into a large mixing bowl. Pour the lemon juice over the burghol and set aside.

Next, finely chop the tomatoes and mint.

Combine all the ingredients in the large mixing bowl, slowly adding the olive oil. Season with salt and pepper to taste. Serve immediately.

Lebanese like to scoop up mouthfuls of the salad with grape, lettuce or cabbage leaves.

Serves 2

Peasant salad

Fattoush

A popular salad that relies on the lemony taste of sumac for flavour and crunchy pieces of Arabic bread for texture.

1 small head of romaine lettuce
1/4 cup onions, diced
1 large tomato, diced
1/4 cup fresh mint, chopped
1/2 cup bekleh (purslane) leaves
1/4 cup parsley, chopped
1/4 cup black olives, pitted
2 large cucumbers, sliced
1 tablespoon dried mint
1/2 round of Arabic bread, fried or toasted and broken into pieces
Salt

Dressing:
2 cloves garlic, crushed
1/4 teaspoon sumac
2 tablespoons lemon juice
1/2 cup olive oil
1 tablespoon of vinegar

Chop lettuce into 1-inch strips, then dice the tomatoes and onions. Next, finely slice the cucumbers and chop the fresh mint and parsley. Remove the stalks from purslane, but keep the leaves whole.

Prepare the dressing by mixing the crushed garlic cloves, lemon juice, vinegar and olive oil in a small bowl.

When ready to serve, combine all the ingredients in a large bowl, adding some of the toasted or fried bread broken into small pieces. Toss with the dressing. Sprinkle with the remaining pieces of bread and sumac.

Serve immediately to keep the bread from becoming soggy.

Serves 4 to 6

Whole bean chickpea salad

Balila

Chickpeas are extremely nutritious and are one of the staples of the Lebanese diet.

2 cups chickpeas, cooked (see page 93)
1 teaspoon ground cumin
Juice of 2 lemons
2 cloves garlic, crushed
Olive oil
Salt

In a large bowl, mix chickpeas while still warm with the ground cumin, lemon juice, salt and crushed garlic. Transfer the mixture to a small bowl or bowls and garnish with olive oil.

This dish is best served warm with Arabic bread.

Serves 2 to 4

Chickpea salad with tomatoes & scallions

Salata bi hommos

This dish can be served as either a starter or a salad.

1 1/2 cup chickpeas, cooked (see page 93)
1 clove garlic, crushed
1/4 cup scallions, diced
1/2 cup tomatoes, diced
3/4 cup lemon juice
1/2 cup olive oil
Salt
Chopped parsley to garnish

In a large bowl, combine the cooked chickpeas, diced scallions, crushed garlic, chopped tomatoes, lemon juice and olive oil. Add salt to taste.

Serve at room temperature. Garnish with chopped parsley.

Serves 2 to 4

Fava bean salad

Salatet foul

A tasty salad, bursting with flavour.

1 cup foul, cooked and chilled (see page 96)
1/4 cup lemon juice
1/4 cup onions, chopped
1/2 cup tomatoes, diced
1/4 cup olive oil
1 clove garlic, crushed
1/4 cup parsley, chopped
Salt

When ready to serve, combine the chilled foul, diced tomatoes, diced onion, chopped parsley and crushed garlic in a large wooden salad bowl, and mix with lemon juice and olive oil. Add salt to taste.

Serves 2 to 4

Eggplant salad

Salatet batinjen

Flavoursome and healthy, an eggplant salad makes a perfect accompaniment to grilled meat.

1 medium eggplant
1 medium tomato
1/2 cup fresh green pepper, chopped
1 garlic clove, crushed
1/2 cup onions, diced
1/4 cup parsley, chopped
1/2 cup olive oil
Salt

Pierce the skin of the eggplant and grill over charcoal or on the top burner of the stove (see page 94) until soft. Peel and mash the flesh. Set aside.

Dice the tomato, onion and green pepper. Chop the parsley and crush the garlic. When ready to serve, mix the ingredients together in a large bowl with olive oil, and add salt to taste.

Serves 2 to 4

Beet salad

Salatet chamandar

An ideal way to add colour to a mezze table, and a good partner for any meat dish.

500 grams medium beets
3/4 cup onions diced
1/2 cup olive oil

Place beets in a saucepan and with plenty of water, and boil for about an hour (to reduce cooking time, use a pressure cooker). Drain and wait for the beets to cool, then peel and dice.

Toss the beets with diced onions and olive oil.

Serve on a small plate.

Serves 2 to 4

Meat

Barbecued lamb marinated in lemon juice & olive oil

Lahem mishwi

During July and August, Beirut residents escape the summer heat and head to the mountains. Barbecuing is a favorite pastime - a way to combine a delicious meal with the great outdoors.

500 grams lamb, cut into bite-sized cubes
3 medium potatoes, sliced
250 grams small onions

Marinade:
1/2 cup lemon juice
1/2 cup olive oil
Salt and pepper

Mix the lemon juice and olive oil together in a small bowl, adding salt and pepper to taste. Marinate meat in the mixture for at least an hour.

Meanwhile, wash and peel the potatoes. Cut them into 1-inch thick slices, and coat with olive oil on both sides.

Thread the marinated meat on skewers, interspersed with small onions. Make sure to leave space between the cubes so that the meat cooks quickly. Grill on hot charcoal for a few minutes, then flip the skewer and repeat for the other side. Once cooked, remove the meat from the skewers and place on a bed of chopped parsley and diced onions, and cover them with Arabic bread to keep warm.

Grill the olive oil-coated potato slices on both sides, and serve as an accompaniment to the meat. Eat with hommos (see page 93) and a tossed salad.

Serves 2 to 4 (makes approximately 8 skewers)

Chicken

Chicken with potatoes, lemon juice & garlic

Dajaj bi batata

A favourite family meal. The potatoes take on a delicious flavour as they soak up the lemon juice and garlic.

1 large chicken
4 potatoes

Marinade:
1 cup olive oil
1/2 cup lemon juice
3 cloves garlic, crushed
Salt

Preheat oven to 400° F (200° C, gas mark 6).

Wash and cut chicken into 4-6 pieces and place in a 2-quart baking dish. Combine olive oil, lemon juice, salt and crushed garlic, and pour the mixture over the chicken. Place the baking dish in the refrigerator, leaving the chicken to marinate for about 2 hours. When ready to dine, wash, peel and cut the potatoes into 3-inch pieces and place in the dish with the chicken. Add a small amount of water, and bake the chicken and potatoes in the sauce for about an hour in a standard oven at 400° F (200° C, gas mark 6), or until golden brown.

Serve hot with a salad.

Serves 4

Barbecued chicken

Dajaj mishwi

Quick and easy: throw it on the barbecue for a summer picnic.

1 large chicken

Marinade:
3 cloves garlic, crushed
1/2 cup lemon juice
1/2 cup olive oil
Salt

Wash and cut chicken into 4-6 pieces. Combine olive oil, lemon juice, salt and crushed garlic, and pour the mixture over the chicken, leaving it to marinate for at least 1 hour. When ready to eat, grill on an open charcoal fire.

Serve the chicken with a dish of hommos (see page 93) and a salad.

Serves 4

Fish

Baked fish

Samak mishwi

Fish are abundant along Lebanon's Mediterranean shores: the local catch can be bought from street vendors along the Corniche, Beirut's coastal promenade. A simple baked fish is traditionally served with a creamy sauce called tarator.

1 medium fish, gutted, cleaned and scaled - Red Snapper or Bass (Lookos)
1 onion, sliced
1/2 cup olive oil
1 lemon, sliced
Lemon wedges
Salt

Preheat oven to 350° F (180° C, gas mark 4).

Rub the inner cavity of the fish with olive oil, then smear olive oil on the outside of the fish as well. Add salt to taste. Inside the cavity, place lemon wedges covered with some of the onion slices and a little more olive oil. Arrange the remaining onion and lemon slices on top of the fish.

Cover the whole fish with foil and bake in a standard oven at 350° F (180° C, gas mark 4), or until soft to touch, approximately 20-30 minutes. Remove foil and bake for an additional 10 minutes.

Serve with tarator (see page 132) and lemon wedges.

or

Hot Pepper Mixture:
1/2 to 1 tablespoon hot pepper powder or paste (*harissa*)
1/2 cup coriander, chopped
2 cloves garlic, crushed
1/2 cup olive oil

Combine hot pepper, chopped coriander, olive oil and crushed garlic, and place the mixture in the cavity of the fish. Top the fish with lemon wedges and sliced onions.

Cover with foil and bake in a standard oven at 350° F (180° C, gas mark 4), or until the fish is soft to touch, approximately 20-30 minutes. Remove foil and bake for an additional 10 minutes.

Serve with tarator (see page 132) and lemon wedges.

Serves 4

Tahineh sauce

Tarator

Sesame oil sauce traditionally served with baked or fried fish.

1/4 cup tahineh
1/2 cup lemon juice
2 cloves garlic, crushed
Salt to taste

Whisk the tahineh, lemon juice and crushed garlic in a bowl. Season with salt to taste. Mix in small amounts of water until the sauce reaches a smooth, creamy consistency.

Serve in a small bowl to accompany baked or fried fish.

Makes approximately 1 cup

Fried fish

Samak mi'li

A quick way of preparing fish when you are in a hurry.

4 small fish, or 1 large fish cut into 4 portions
1 cup or more of flour
Salt to taste
Olive oil for frying
Arabic bread

Wash, clean and scale fish. Dredge in flour and fry in olive oil.

After removing the fried fish from the skillet, divide a round of Arabic bread in quarters and fry in the same oil.

Serve fish hot with tarator (see page 132), lemon wedges and the fried Arabic bread.

Serves 2 to 4

Vegetarian

Eggplant stew

Mousaka'a

Not to be confused with the well-known Greek dish of the same name, Lebanese mousaka'a is vegetarian and consists of layers of eggplant, cooked gently in a tasty tomato sauce.

2 medium eggplants
2 medium tomatoes, diced
1 large onion, diced
2 tablespoons tomato paste
1/2 cup of chickpeas, cooked (optional) (see page 93)
Olive oil for frying
Salt

Peel and thinly slice the eggplant. Salt and set aside for at least 1 hour. (Salting the eggplant first reduces the amount of oil absorbed while frying, and draws out the bitter juices.)

Meanwhile, dice the onions and tomatoes. Rinse off the eggplant slices, pat them dry and fry in olive oil until golden brown.

Place a layer of diced tomatoes at the bottom of a large saucepan. Next, add a layer of eggplant slices on top of the tomatoes, and a layer of diced onion on top of the eggplant. Repeat this pattern at least one more time.

Dilute the tomato paste in 2 1/2 cups of water and pour over the layered vegetables in the saucepan. Cook on low heat for approximately 15 minutes, or until onions are cooked and the sauce has thickened.

Serve at room temperature.

I do not cook this dish with chickpeas. However, some people like to include a fourth layer of cooked chickpeas along with the tomatoes, eggplant, and onions. Simply add a chickpea layer into the rotation, and continue to follow the directions as described above.

Serves 2 to 4

Stuffed grape leaves

Warak ainab bi zeit

Rolling grape leaves is a time-honoured tradition in Lebanese homes, and often a communal activity. You can also use the stuffing to make rolled Swiss chard leaves, stuffed squash or stuffed eggplant.

Meatless stuffing for all vegetables:
1/2 cup rice, uncooked
1 onion, diced
3 cups parsley, chopped
1/4 cup chickpeas, cooked (see page 93)
1 cup tomatoes, diced
1/2 cup lemon juice
1/2 cup olive oil
Salt

(Mix the above ingredients together to make the filling)

Fresh grape leaves or preserved grape leaves (approximately 50)

To prepare fresh grape leaves, dip them in hot water and drain. If using preserved grape leaves, rinse them beforehand to remove some of the salty flavour.

Place the shiny side of the leaf down on the table, so the coarse side faces upwards. Put about 1 tablespoon of stuffing along the top of edge of the leaf, leaving a 1/2 inch border along the top and sides. Fold the top part of the leaf over the filling. Then, fold the sides over the filling (towards the centre of the leaf). Roll the rest of the leaf downwards towards the tip end of the leaf to make it into a cigar-like shape. The roll should be snug but not too tight, as the filling will expand while cooking.

Place rolls side-by-side at the bottom of a saucepan, with the folded side of the rolls facing down so they won't come undone. Add salted water (just more than enough to cover all the grape leaves), and add a little more lemon juice and olive oil. Cover with an inverted plate, and cook on low heat for about 30 minutes or until the rice is cooked.

Serve hot or at room temperature.

Makes approximately four dozen medium-sized grape leaves

Lentils and rice

Moudarddarah

Moudarddarah, although the epitome of simplicity, is an immensely popular dish in Lebanon.

1 cup brown lentils
3 onions, thinly sliced
1/2 cup rice
1/4 cup olive oil
Salt

Spread the lentils on a tray to check for any pieces of grit. Rinse lentils.

Rinse the rice until the water runs clear; drain, and set aside. Place the lentils in a saucepan and cover with water. Bring to a boil, then reduce to moderate heat and cook for about 30 minutes, or until soft. Place the rice in the saucepan with the cooked lentils, adding salt and a little more water. Simmer for about 20 minutes, or until rice is cooked and all the water has been absorbed.

Place the thinly-sliced onions in a skillet, and fry in olive oil until golden brown. Mix some of the fried onions and the olive oil into the lentil mixture.

Transfer moudarddarah to a platter and garnish with the rest of the fried onions.

Serves 4

Mashed lentils and rice

Moujaddarah

Lentils, packed full of vitamins and dietary fibre, make an economical meat substitute.

1 cup brown lentils
1/2 cup rice
3 onions, diced
1/2 cup olive oil
Salt
Scallions to garnish

Spread the lentils on a tray to check for any pieces of grit. Rinse lentils and cook in a saucepan covered with water for about 30 minutes.

Rinse the rice until the water runs clear, and drain. Place rice in another saucepan, and cook in water for 20 minutes, or until all the water is absorbed.

Fry the diced onions until golden brown, and add to the cooked lentils. Place the mixture in a blender and pulse to make a smooth paste. Return the mixture to the saucepan, add the cooked rice and simmer for at least 10 minutes, stirring regularly.

Serve on individual plates at room temperature, and garnish with scallions.

Serves 4

Mixed vegetables fried in olive oil

Khudra makliah bi zeit

A delightful assortment of vegetables best eaten with fattoush.

2 large potatoes, peeled and sliced
1 medium eggplant, peeled and sliced
3 medium squash, sliced lengthwise
2 medium peppers, sliced lengthwise
1 medium cauliflower, separated into florets
Olive oil for frying
Salt
Fresh mint, scallions and radishes to garnish

Peel, slice and salt the eggplant, and set aside for an hour. (Salting the eggplant first reduces the amount of oil absorbed while frying and also removes the bitter juices.)

When you are ready to cook, prepare the other vegetables for frying. Rinse eggplant and pat dry with kitchen towel.

Heat 1 cup of olive oil in a frying pan. Fry each vegetable in turn, adding more oil as necessary. Remove each vegetable as it is fried and place it on a paper towel to help drain excess oil. Keep the squash for last, as oil darkens its colour when fried. Discard the oil.

Place all the fried vegetables on a platter and garnish with fresh mint, radishes, scallions and a fattoush salad (see page 111).

Serves 4 to 6

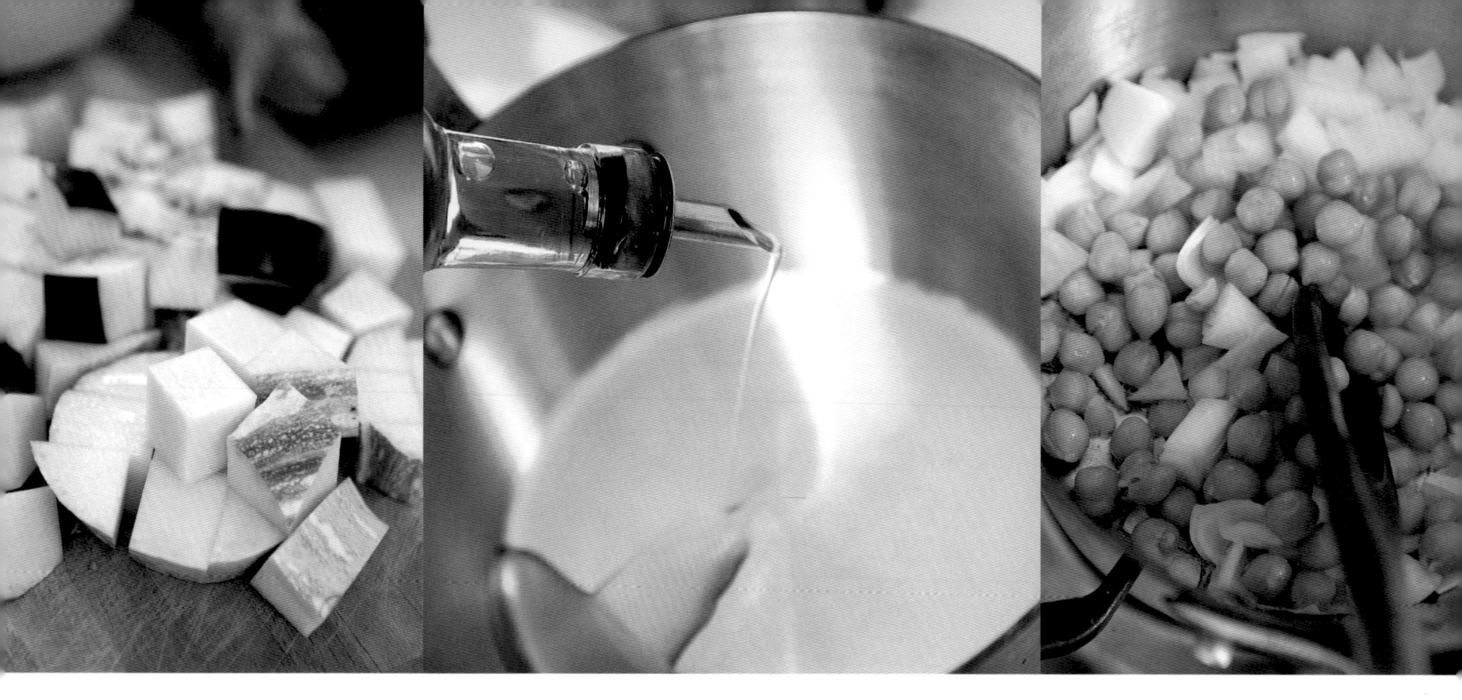

Squash and eggplant stew

Yakhnet al batinjen w koussa

An eastern cousin of the French Provençale dish, Ratatouille.

250 grams squash
250 grams eggplant
2 cloves garlic, crushed
1/2 cup of chickpeas, cooked (see page 93)
1 onion
1 tomato, chopped
1 heaping tablespoon tomato paste
1 heaping tablespoon dried mint
1/4 cup olive oil
Salt and pepper

Peel and dice the eggplant and the squash into 2-inch cubes and set aside. In a large saucepan, heat the olive oil and sauté the chopped onions and chickpeas. Next, add the cubed vegetables. When the mixture is well coated with oil and limp, add the chopped tomato.

Dilute the tomato paste in 3 cups of lukewarm water and add to the vegetables. Cover and bring to a boil, then simmer on medium heat for approximately 30 minutes, or until all the vegetables are soft to touch. Add the dried mint and crushed garlic and simmer for an additional 10 minutes. Salt and pepper to taste.

The stew should be served with rice, hot or at room temperature.

Serves 4

Okra stew

Yakhnet al bamieh

A mellow stew. Try to use young okra, as the older pods tend to get tough and stringy.

500 grams fresh okra
2 tablespoons tomato paste
2 cloves garlic, crushed
1/2 cup coriander, chopped
2 tomatoes, diced
1 large onion, diced
1/2 tablespoon dried coriander
1 tablespoon dibs remmane
1/2 cup olive oil
Salt and pepper

Rinse the okra and pat dry. Trim the stems with a knife. Cooked okra has a tendency to be slimy: in order to prevent this, cut the stem in a cone shape, and avoid puncturing the pod.

Heat the olive oil and fry okra until limp. Remove and set aside. In the same oil, lightly sauté the diced onions. Add more oil as needed. Remove the onions and, still using the same oil, sauté the chopped coriander and crushed garlic.

Place the okra, diced onions, coriander mixture and tomatoes in a 2-quart saucepan. Dilute the tomato paste with 3 cups of lukewarm water, and pour over the stew until it is covered. Add salt and pepper to taste. Cover the pot and bring to a boil, then simmer for about 30 minutes. Add the dibs remmane and cook for an additional 10 minutes.

Serve with rice, hot or at room temperature.

Serves 2 to 4

Spinach in olive oil

Sabaneikh bi zeit

An excellent vegetarian option, spinach is rich in folic acid, vitamins C and E and iron.

500 grams fresh spinach, chopped
1 large onion, thinly sliced
2 cloves garlic, crushed
1/4 cup olive oil
Salt

Wash the spinach and cut into 2-inch strips, then boil for a few minutes in salted water. Drain, squeezing out any excess water by hand. Allow the spinach to cool, then mix in the crushed garlic and set aside.

Fry the thinly-sliced onion in hot olive oil until golden brown. Mix the oil and some of the onions into the spinach mixture. Salt to taste. Garnish with the rest of the onions.

Serve at room temperature with lemon wedges.

Serves 2 to 4

Potato pie

Kibbet batata

A substantial, hearty meal, ideal for a winter lunch or dinner.

2 large potatoes
1 cup burghol (cracked wheat)
1 large onion, chopped
1 tablespoon onion, grated
1/2 cup walnuts, chopped
1/2 teaspoon dibs remmane
1/2 cup olive oil
Salt

Preheat oven to 400° F (200° C, gas mark 6).

Rinse the burghol until the water runs clear. Leave to soak in water for an hour.

Boil the potatoes for 15-20 minutes, depending on the variety. Peel and mash the potatoes in a blender or food processor. Add the grated onion to the potato mash and pulse until blended. Set aside.

When ready to make the pie, drain the water from the burghol, squeezing out any excess water by hand. Add the burghol to the potato and onion mash in the blender and pulse until the mixture is smooth. Season with salt and pepper to taste.

Fry the chopped onions in hot olive oil until they are translucent. Add the walnuts to the skillet briefly, just long enough to coat them with oil - about 1 minute. Remove from heat and stir in the dibs remmane.

Grease a 22-centimetre round glass baking dish with olive oil, and divide the potato mixture into two portions. Flatten the first portion in the dish to make a smooth bottom layer and cover with the walnut, onion and dibs remmane mixture. Then flatten the other half of mashed potatoes on top. Score the surface with a knife in a diamond shape design, and drizzle with 2 tablespoons of olive oil. Bake in a preheated oven at 400° F (200° C, gas mark 6) for 30 minutes, or until the top becomes golden brown.

Serve at room temperature with a salad.

Serves 4

appendix

PLACES OF INTEREST

History
Shebaa Mills
AUB Museum
National Museum

Ancient olive trees
Bshaale, near Douma
Ramlieh village, near Aley
Trees planted by Emir Fakhreddine on the outskirts of Baakline village

Soap making
Audi Soap Museum, Sidon
Zahia Abboud: for soap production and sculpture, Sidon
Hassoun soap, Tripoli

Archaeological sites & ancient olive oil presses
Byblos, near the Crusader castle
Shim in the Shouf Mountains, near the Shamoun Palace
Khalde, on the southern outskirts of Beirut
Oumm el-Amed, near Tyre
Deir el-Kalaa, near Beit Mery

Mills to visit during pressing
For the centrifugal method, visit one of several in Hasbaya
For a traditional stone mill, visit Zahia Abboud's press in Sidon
For a modern continuous decanter, visit Kfifane press in the Batroun region
(See glossary of producers)

Tasting session
Visit Tanail for a tasting session with Ramzi Ghosn, proprietor of Massaya wines
Visit Ba ino for a tasting session with Youssef Fares, agricultural engineer and owner of Olive Trade

(See glossary for list of all producers and contact details)

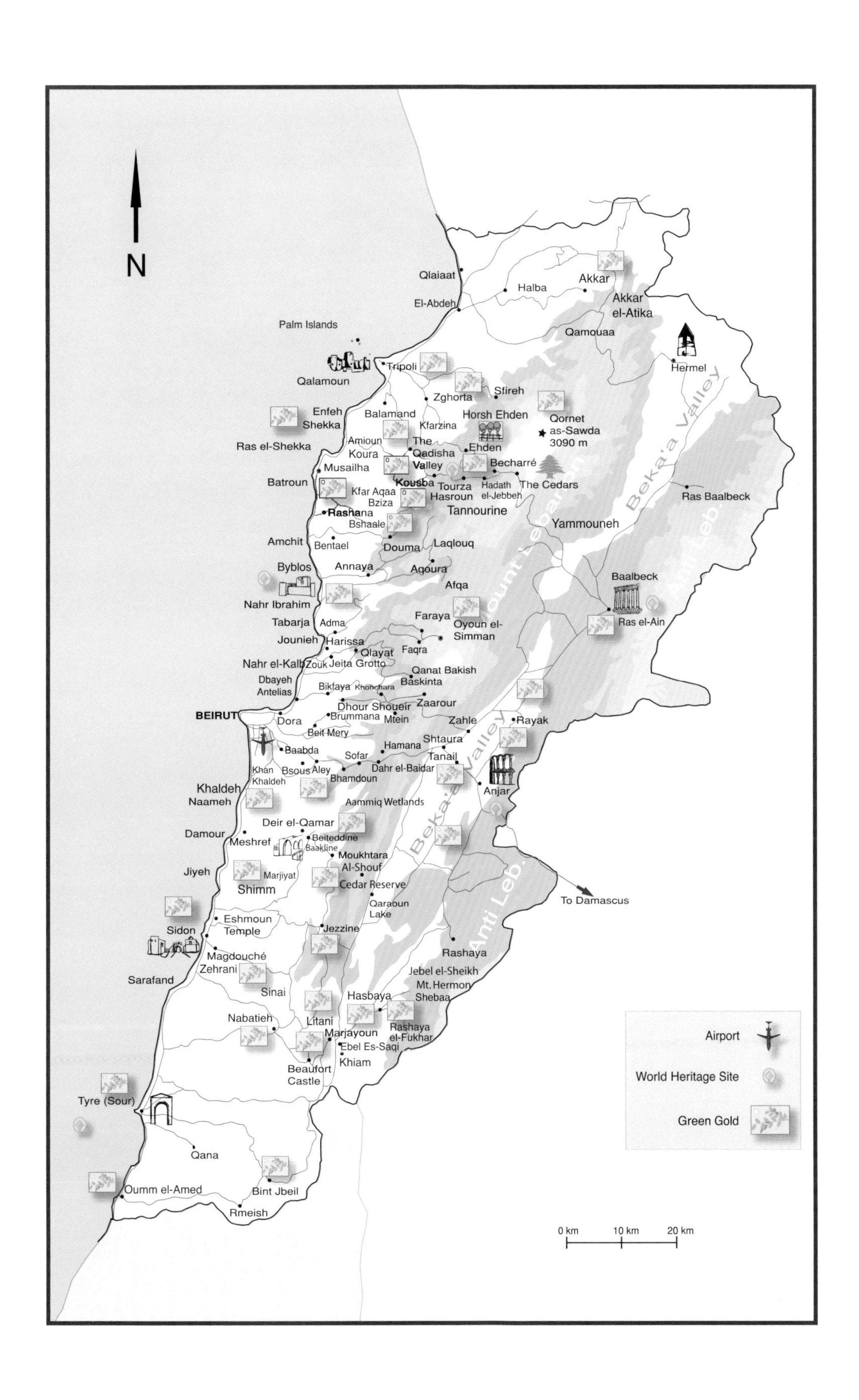
N
Qlaiaat
Halba
Akkar
El-Abdeh
Akkar el-Atika
Palm Islands
Qamouaa
Tripoli
Hermel
Qalamoun
Sfireh
Zghorta
Enfeh
Shekka
Balamand
Horsh Ehden
Qornet as-Sawda 3090 m
Kfarzina
Ras el-Shekka
Amioun
Koura
The Qadisha Valley
Ehden
Musailha
Becharré
Batroun
Kousba
Tourza
Hadath el-Jebbeh
The Cedars
Kfar Aqaa
Hasroun
Bziza
Tannourine
Ras Baalbeck
Rashana
Bshaale
Yammouneh
Amchit
Bentael
Douma
Laqlouq
Byblos
Annaya
Agoura
Mount Lebanon
Beka'a Valley
Anti Leb.
Afqa
Baalbeck
Nahr Ibrahim
Faraya
Tabarja
Adma
Oyoun el-Simman
Ras el-Ain
Jounieh
Harissa
Faqra
Qlayat
Nahr el-Kalb
Zouk
Jeita Grotto
Qanat Bakish
Dbayeh
Baskinta
Antelias
Bikfaya
Khonchara
Dhour Shoueir
Zaarour
BEIRUT
Brummana
Mtein
Zahle
Rayak
Dora
Beit Mery
Shtaura
Hamana
Baabda
Sofar
Tanail
Khan Khaldeh
Bsous
Aley
Dahr el-Baidar
Bhamdoun
Anjar
Khaldeh
Naameh
Aammiq Wetlands
Deir el-Qamar
Damour
Meshref
Beiteddine
Baakline
Moukhtara
Al-Shouf
Jiyeh
Marjiyat
Shimm
Cedar Reserve
Qaraoun Lake
To Damascus
Eshmoun Temple
Sidon
Jezzine
Rashaya
Magdouché
Zehrani
Jebel el-Sheikh
Sarafand
Mt. Hermon
Sinai
Shebaa
Hasbaya
Nabatieh
Litani
Rashaya el-Fukhar
Marjayoun
Ebel Es-Saqi
Khiam
Beaufort Castle
Tyre (Sour)
Qana
Bint Jbeil
Oumm el-Amed
Rmeish
Airport
World Heritage Site
Green Gold
0 km
10 km
20 km

Olive Oil Distributors in Lebanon

DISTRIBUTOR	ADDRESS	CONTACTS
Al Dayaa Products www.aldayaa.com	P.O. Box 12 Amioun Al Koura Kafer Akka	Fadi Tannous Tel: 06-952 838, 03-303 862 Email: aldayaa@hotmail.com
Al Rabih Saifan Press www.alrabih.com.lb	P.O. Box 70070 Dbayeh	Georges Nasraoui Tel: 04-541 207, 03-543 737 Fax: 04-541 209 Email: sonaco@alrabih.com.lb
Al Wadi Al Akhdar www.alwadi-alakhdar.com	Amaret Shalhoub, Zalka P.O. Box 11-2652, Beirut	Gay Mandour Tel: 01-892 035 Fax: 01-898 974 Email: info@alwadi-alakhdar.com
Boulos Olive Oil, ATYAB www.zeitboulos.com	ATYAB P.O. Box 1646 Jounieh	Tony Maroun Tel: 09-918 525 Fax: 09-910 090 Email: atyab@atyab-lb.com, Info@zeitboulos.com
Cortas Oil www.cortasfood.com	Cortas Building Tripoli Road Dora, Beirut	Tel: 01-257 171 Fax: 01-257 272 Email: sales@cortas.com.lb
Gardenia www.gardeniaspices.com	Ksara Gardenia Center	Nicola Abou Fayssal Tel/Fax: 08-814 897, 08-818 661 Email: info@gardeniaspices.com
Khater Brothers	Hamad Building, Mama street Tallat el-Khayyat, Beirut	Tel: 01-815 136, 03-828 562
Wadi el Kheir Foods www.wadielkheirfoods.com	Al Mouhaidatha, Rachaya el-Wadi	Fadi Jammal Tel: 08-591 383, 03-231 951 Fax: 08-561 555 Email: info@wadielkheirfoods.com
Zeytouna www.zeytouna.com	Kfifane - Batroun	Roland Aindari Tel: 06-720 078, 03-590 920 Email: zeytouna@zeytouna.com
Second House Products www.secondhouseprod.com	Mazraat Yachouh Industial zone, E Street, Mtein	Francois Rizk Tel: 04-915 391

*To dial from abroad, drop the leading zero and add your country's international access code, followed by Lebanon country code 961.

Olive Oil Producers in Lebanon

PRODUCER	BRAND	ORIGIN	TELEPHONE	
Habib Mitri	Salihiyye Oil	Salhiyye	07-230 560	03-266 683
Joseph Khoury	Hiram	Zghorta	03-228 849	06-550 450
Vicky Khoury	Rose de Tyr	Rishmaya, Aley	03-246 526	01-447 960
Nizar Mokhder	Mira	Babliyye	03-420 222	07-420 333
Joe Abou Rjaili	Les Olivades	Habramoun	03-302 317	
Moussa Ghantous	Al Khalil	Koura	03-437 178	
Ghaith Maalouf	Rashaya el-Fukhar	Rashaya el-Fukhar	03-750 470	
RMF (René Moawad Foundation)	Zeytouna	Kfifane	03-590 920	
Gaby Beainy	Gaby Beainy	Bkassine	03-743 439	
Sleiman el Hage	Sleiman el Hage	Aaramta	03-838 162	
Nabil Mohammad Kobaiter	Al Wazir Oil	Rasheen	03-295 225	
Antoine Lakis	Al Balad	Koura	03-686 644	
Simon Atieh	Pronto	Koura	03-800 096	
Tony Maroun	Boulos Olive Oil	Koura	03-629 429	
Yehya Hassan		Aaba, Nabatieh	03-428 693	
Samir Qaaqour		Baasir, Shouf	07-920 100	
Amin Ibrahim	AgriCoop Association	Bnehran, Koura	03-813 480	
Ismail Amin	Hasbaya Agricultural Centre	Hasbaya	03-367 267	
Kamal Faran		Mazret Bayad	03-241 757	
Lamia Rassi	Fondation Abdallah Rassi	Sheikh Taba, Akkar	03-223 887	
Youssef Fares	Zejd - Olive Trade	Baino, Akkar	03-283 724	
World Vision	Campagnia Olive Oil	Marjayoun	04-401 980	
Zahia Abboud	Zahia Abboud	Sidon	07-733 676	
La Route de l'Olivier	Healthy Basket	Beirut	03-733 227	

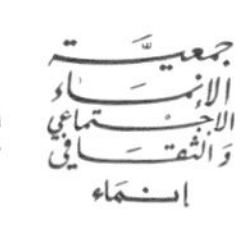

Soap Producers

Zahia Abboud
Tel: 07-733 676
Email: zahia_abboud@hotmail.com

Khan Al-Saboun Hassoun
Tel: 06-438 369
www.khanalsaboun.com

NGOs and Associations

World Vision Lebanon
Tel: 04-401 980
www.lebanon.worldvision.org

SRI International (Stanford Research Institute)
Tel: 01-983 008
www.inmadevelopment.org

Mercy Corps
Tel: 01-383 919
www.mercycorps.org

ICU (Institute for University Cooperation)
Tel: 07-840 047
icusud@cyberia.net.lb

RMF (René Moawad Foundation)
Tel: 01-613 367
rmf@rmf.og.lb

Souk el Tayeb
Tel: 03-340 198
www.soukeltayeb.com

Healthy Basket
Tel: 03-733 227

Bibliography

Amouretti, M.-C. & Comet G., *Le Livre de l'Olivier*, édisud, 1985.

Callot, O., *Les Huileries antiques de la Syrie du Nord*, Librarie Orientaliste, Paul Geuthner, 1984.

Chanesaz, Moheb Nader, *Les pressoirs d'olive au Liban: changements techniques et socioculturels. Recherches de terrain en technologie culturelle*, Thèse de doctorat soutenue a l'Université Lumière-Lyon 2, 2004.

Chanesaz, Moheb Nader, Sweet & Bitter, *ARAM Volume 17*, 2005.

De Barry, Nicolas, *L'ABCdaire de l'huile d'olive*, Flammarion, 1999.

Divine Islam's Koran Viewer software, v2.9

Hadjisavvas, Sophocles, *Olive oil processing in Cyprus from the Bronze Age to the Byzantine Period*, Paul Astroms Forlag, 1992.

Holy Bible: *New International Version*, Hodder and Stoughton, 1978.

Homer, *The Iliad*, translated by E.V. Rieu, Penguin Classics, 1950.

Impact Assessment of the Association Agreement on the Olive Oil Sector in Lebanon, Ministry of Economy and Trade, 2006.

Ismail, Adel, *Lebanon*: *History of a People*, Dar Al-Makshouf, 1972.

'Mishwar' - Promenade Series, INMA, SRI, USAID and Ministry of Tourism, 2005.

Thubron, Colin, *The Hills of Adonis*, William Heinemann, 1968.

Waliszewski, Tomasz et Ortali-Tarazi Renata, *Village Romaine et Byzantine à Chhim-Marjiyat*, *BAAL 6*, 2002.

www.lebaneseoliveoil.com

Acknowledgements

So many have contributed to the writing of this book, from the farmers who so patiently answered my questions to the village elders who appeased my endless search for usage tips and even proverbs, that it is impossible to thank them all. My precious family, Henri,Tara and William, who humoured my passion and nurtured it - suffice it to say that without their contributions, this book and my journey would be incomplete.

Firstly, my gratitude is extended to my publisher Charlotte Hamaoui for giving me the opportunity to write this book. Even throughout the tumultuous summer of 2006, she was able to see her vision flourish from concept through to production. Thank you to Eleena Sarkissian for her diligence and patience in helping the flow towards completion. Thanks also to Mohammed and Hassan for their driving skills along the sinuous roads of Lebanon's mountains.

I also wish to thank the brilliant Roger Moukarzel and his assistant Steve Kozman for the spectacular photographs that have made this book a visual treat, and the wonderful Betty Sabieh for sharing her tried and tested recipes. Also Kamal Mouzawak, who provided fabulous food styling and pictures of Lebanon's culinary delights; he remains tireless in his efforts to preserve local gastronomic traditions in a country where food is taken very seriously indeed.

A big thanks to my astute family practitioner, Pierre Awad, MD, for his advice on the medical uses of olive oil. Also to my many supportive friends who offered advice - and so obligingly tested the beauty tips!

James Billings, (director, SRI International, Lebanon office), deserves recognition for his efforts to promote Lebanese olive oil worldwide, as does Nell Abou-Ghazale, (agriculture coordinator), for organising Lebanon's first olive oil competition at HORECA 2006.

A special show of appreciation goes out to Dr Moheb Chanesaz, professor of anthropology, at the Lebanese University, who amazed me over many hours in coffee shops patiently explaining his theories regarding the linguistics related to the extraction of olive oil. Thanks to Sheikh Sleiman El Dagher, president of SILO (Syndicate of Interprofessional Lebanese Olive Oil Producers) for his invaluable information regarding the local industry. Also Thana Rida Abu Ghyda, agronomist, Ministry of Economy and Trade, for some helpful facts and figures and Nabil Moawad, head of agriculture department at the René Moawad Foundation, for his obliging information regarding the olive oil market and his open invitation for me to join industry lectures.

Along the way, the name of Hussein Y. Hoteit, agricultural engineer, came up wherever I went, as the expert to talk to. He gave me invaluable advice regarding cultivation practices and innovations. And hats off to Youssef Fares, agricultural engineer and manager of Olive Trade, for his foresight into organic farming practices and for his patient response to my countless queries. He is both diligent and judicious beyond his years - and he produces delicious olive oil!

Thank you to Father Ksas for confirming and adding to my biblical citations and to Dr Mohammed Y. Najm, professor Emeritus of Arabic, American University of Beirut for the Koran references. Assaad Seif, archaeologist, DGA (Directorate General of Antiquities), provided me with various references regarding olive oil in antiquity. I also extend my thanks to the accommodating folk at World Vision International, Team South who shared with me their knowledge of organic farming practises and their hopes for the farmers of the South.

I am forever indebted to Dr Rami Zurayk, professor of rural agriculture, American University of Beirut, who guided my journey and shared his passion for his native terra firma with me. His vision for Lebanon's rural regions is inspirational – and he makes stunning olive oil soap.

Finally, I need to give a special mention to my dear friend Rodney Wallace, a management consultant and talented musician (and recent convert to olive oil soap!), whose only connection to Lebanon is through me. Whilst putting the finishing touches on the book during a visit to Austria in July 2006, I was devastated to hear the news that Lebanon was once again at war. Subsequently, my thoughts returned to those at home and the book took second place. During these troubled times, Rodney went over the manuscript with me prior to publication and helped restore my confidence to proceed with its completion. In turn, I hope this book goes a little way to provide inspiration for all who read it and for the people of Lebanon to whom it is ultimately dedicated.

- Sabina, August 2006

index